LHMC BOOK

54008000001253

ST BARTHOLOMEW'S AND THE ROYAL LONDON
SCHOOL OF MEDICINE AND DENTISTRY
WHITECHAPEL LIBRARY, TURNER STREET, LONDON E1 2AD
020 7882 7110

4 WEEK LOAN
Books are to be returned on or before the last date below,
otherwise fines may be charged.

12 DEC 2003		
12 JAN 2004		
04/06/04		

L. H. M. C.
LIBRARY
ACCESSION No.

Introduction to Problem Solving in Biomechanics

ɑ 19-59

HKL 0166612 20/12/88

WITHDRAWN
FROM STOCK
QMUL LIBRARY

Introduction to Problem Solving in Biomechanics

Christina von Heijne Wiktorin, R.P.T., M.A.
Industrial Physical Therapist
Swedish Construction Industry's Organization of Occupational
Health and Safety (Bygghälsan)
Karlstad, Sweden

Margareta Nordin, Ph.D.
Associate Director
Occupational and Industrial Orthopaedic Center
Hospital for Joint Diseases Orthopaedic Institute;
Assistant Adjunct Professor
Program Director of Ergonomics and Occupational Biomechanics
Department of Occupational Health and Safety
New York University
New York, New York, U.S.A.

Kajsa Forssén, Illustrator
Marianne Westerdahl, R.P.T., Illustrator
Laurie Glass, Editorial Associate

Lea & Febiger   Philadelphia
1986

Lea & Febiger
600 Washington Square
Philadelphia, PA 19106-4198
U.S.A.
(215) 922-1330

WE103   WIK

Library of Congress Cataloging in Publication Data

Wiktorin, Christina v. Heijne.
  Introduction to problem solving in biomechanics.

  Translation of: Exempelsamling i biomekanik.
  Bibliography: p.
  1. Human mechanics—Problems, exercises, etc.
2. Physical therapy—Problems, exercises, etc.
3. Biomechanics—Problems, exercises, etc. I. Nordin,
Margareta. II. Title
QP303.W5713   1985   615.8′24   84-26076
ISBN 0-8121-0941-4

L. H. M. C.
LIBRARY
ACCESSION No.
49519

Copyright © Christina v. Heijne Wiktorin, Margareta Nordin and
Studentlitteratur ab, Lund, Sweden, 1982

Copyright © 1986 by Lea & Febiger. Copyright under the International Copyright Union. All Rights Reserved.
This book is protected by copyright. No part of it may be reproduced in any manner or by any means without
written permission of the Publisher.

Printed in the United States of America

Print Number: 5 4 3 2 1

To Anna and Erik

Foreword

Good therapy is based on sound knowledge. The therapist needs a fundamental understanding of the medical problem under treatment, a basic grasp of anatomy and physiology, and, when therapy is applied to the musculoskeletal system, a thorough knowledge of biomechanics.

How to acquire a knowledge of biomechanics has been the subject of intense study over the past 20 years as the field has evolved rapidly. Biomechanics cannot be learned by observation. It cannot be learned by the average therapist through the rigid mathematical approach of engineering. Although didactic lectures play a major role, the principles of adult education dictate that participation in the educational process is a key factor in the absorption of knowledge.

In the case of biomechanics, problem solving offers the best methodology for teaching the student who is not mathematically oriented. By actually working a problem, the student identifies the underlying concepts demonstrated in the problem and integrates those concepts with his or her existing knowledge.

In this volume of clinical examples in biomechanics, therapists Christina v. Heijne Wiktorin and Margareta Nordin have provided a set of problems that are of direct significance to all those who deal with disorders of the musculoskeletal system. By solving the problems in this volume and giving careful attention to the discussion questions and answers, the reader will gain a knowledge of biomechanics that will prove most valuable in daily therapeutic activities.

Victor H. Frankel, M.D., Ph.D.
Director
Department of Orthopaedic Surgery
Hospital for Joint Diseases
Orthopaedic Institute
New York, New York

Foreword

The field of biomechanics combines engineering with biology and physiology. In this field the laws of physics are adapted to describe and explain biological phenomena. Thus, the subject is broad, ranging from the properties of liquids to those of solids, from the static state to the dynamic.

For those of us who deal with the functions, diseases, and injuries of the locomotor system, biomechanics is a fundamental aspect of our work. Our diagnoses and therapy are based in large measure on biomechanical principles.

This collection of clinical examples in biomechanics will be of value to everyone who works with disorders of the musculoskeletal system. It is directed primarily to physical therapists, but it also has an important mission to fulfill for professionals in related fields. Through the practical examples and accompanying problems presented here, we will learn the significance of biomechanics to our work. Of utmost importance in this regard is that we will come to understand the principles that determine loading of the musculoskeletal system, and we will learn to apply these principles in practical work situations.

The problems can be solved with the use of mathematics and physics, but problem solving should not be the reader's only aim. What is more important is that the underlying principles of biomechanics are thoroughly grasped. Furthermore, the book will not "come alive" until the theoretical examples are carried over into practical life. I am therefore delighted to find that the authors have chosen examples that closely resemble situations in the workaday world of treatment.

This book has an important purpose, namely, to stimulate interest in biomechanics and thus to improve the care of patients with disorders of the musculoskeletal system.

Gunnar Andersson, M.D., Ph.D.
Department of Orthopaedic Surgery
Sahlgren Hospital
University of Gothenburg
Gothenburg, Sweden

Preface

The clinical examples in biomechanics presented in this volume are intended primarily for physical therapists but will also be of value to physicians, physical trainers, occupational therapists, osteopaths, and others who deal with rehabilitation of the musculoskeletal system.

The mathematical problems accompanying each example, which are limited to static treatment of rigid body mechanics, are designed to represent actual clinical situations and are suited for use in the teaching of basic biomechanics. Because the book also includes answers to all the problems and more extensive solutions for many of them, it is appropriate for both individual and classroom study. Care has been taken to select problems that illustrate how the laws of mechanics are applied to the locomotor system. In addition, the problems chosen are intended to stimulate discussion on the methodology underlying rehabilitation of the musculoskeletal system.

Background material for this book was collected over many years of teaching experience, partly from the four-year period during which Christina v. Heijne Wiktorin taught at the Physical Therapy School in Gothenburg, Sweden, and partly from several courses on basic biomechanics held by both authors for various professional groups.

Several individuals contributed greatly to making this book a reality. In the fall of 1978, Christina v. Heijne Wiktorin had the privilege of spending several months at the Department of Anatomy at the Karolinska Institute in Stockholm, where she devised some problems under the guidance of Professor Jan Ekholm.

In the fall of 1980 and the winter of 1981, Christina v. Heijne Wiktorin was a fellow in the Occupational and Industrial Orthopaedic Center at Sahlgren Hospital in Gothenburg, where Margareta Nordin was the chief ergonomist. Her stay was made possible through the assistance of Professor and Chairman Alf Nachemson and Docent Gunnar B. Andersson in the Department of Orthopaedic Surgery at Sahlgren Hospital.

Many colleagues in the physical therapy field at Sahlgren Hospital, Gothenburg, and Danderyd Hospital, Stockholm, were very helpful in collecting data and contributing many useful ideas. Registered physical therapist Marianne Svedenkrans provided data and illustrations for example 10.5.

Example 7.7 is the work of Docent Gunnar Andersson, who also examined the manuscript for the Swedish edition of this book with an expert's eye, paying particular attention to the medical aspects of the work, and contributing many valuable suggestions.

Lars Wiktorin, Ph.D., was most helpful in examining the material from a technical standpoint. He reviewed all of the examples and contributed much constructive criticism.

The photographers at Danderyd Hospital and Sahlgren Hospital assisted in producing the photographs used for most of the illustrations. Working with them and with the skillful illustrators, Kajsa Forssén and Marianne Westerdahl, was a great pleasure.

Secretaries Anette Ekwall, Eva Allers-Wikström, and Gunvor Johansson demonstrated endless patience and care in the typing of the Swedish manuscript.

The manuscript for the English version was edited painstakingly by Laurie Glass, who through her command of the language, firm grasp of biomechanics, and dedication to the project greatly enhanced the readability of this version. The English manuscript was word processed with utmost speed and intelligence by Judy Raymond, whose exceptional skills and sparkling enthusiasm made her an important member of our production team. Her husband Victor provided essential technical assistance in the production process.

To those friends and colleagues, and to all the others who in various ways participated in the production of this book, we extend our warmest thanks.

Financial assistance for the project was provided by the Occupational and Industrial Orthopaedic Center of Sahlgren Hospital and by grants from the Dr. Gunnar Svantesson Memorial Fund, the Swedish Association of Registered Physical Therapists, and the Metal Trades Employers Association.

We acknowledge that despite all of the help that we received in checking the material for this volume, some errors and omissions are inevitable. For these we accept full responsibility. Therefore, we will appreciate receiving any comments and suggestions that may in some way improve the contents of future editions of this book.

Christina v. Heijne Wiktorin, Karlstad, Sweden

in cooperation with *Margareta Nordin, New York, U.S.A.*

Contents

Examples in Chapters 4 through 10

All of the examples in Chapters 4 through 10 are categorized below according to their mechanical content to allow the reader to locate examples of greatest interest. Under each heading the examples are listed in order of increasing difficulty.

Resolution of Forces

10.1. Loads through the Chin during Traction Treatment of the Cervical Spine

Composition of Forces

8.6. The Patellofemoral Joint Reaction Force during Quadriceps Muscle Exercises with the Knee Joint Extended and Flexed

Torque

8.1. Change in the Direction of an Externally Applied Force during Quadriceps Muscle Exercises

8.2. The Step Test. A Functional Strength Test of the Knee Muscles

7.1. Stretching the Hamstring Muscles Passively in Two Different Ways

10.2. Methods of Applying a Halter during Treatment with Cervical Traction

8.3. Strength Testing of the Knee Flexor and Extensor Muscles with the Dynamometer

4.1. Simple Heel Raise. Functional Strength Test of the Plantar Flexor Muscles

9.1. The Erector Spinae Muscle Force during Lifting

8.4. Quadriceps Muscle Exercises with a Weight Boot

5.1. Strengthening the Elbow Extensor Muscles with Dumbbells

5.2. Performing Push-ups in the Sitting and Prone Positions

6.1. Strengthening the Shoulder Flexor Muscles with Pulleys

8.5. Exercising the Knee Extensor Muscles in a Sitting Position on the Quadriceps Table

7.2. Active Straight Leg Lifts Performed in the Supine Position

7.3. Analysis of a One-Leg Stance Using Two Free Body Diagrams

Instructions to the Reader

The purpose of this book is to illustrate, by means of clinical examples, how the laws of mechanics can be applied to the human body. In other words, the principles of biomechanics are brought to light. For each example, several problems involving the use of mathematics and physics have been devised. Because the answers to all problems and complete solutions to selected problems are given, the book is appropriate for both classroom instruction and individual study. Solving the problems requires an elementary knowledge of mathematics and physics as well as knowledge of anatomy. The problems need not be solved in sequence; the reader can choose to solve any of them found to be relevant.

Chapter 1   This chapter provides a brief review of the mathematics required for solving the problems in Chapters 4 through 10. Review problems are presented, along with the complete equations for their solution.

Chapter 2   This chapter provides an overview of the basic laws and concepts in biomechanics and can be read as an introduction to Chapters 4 through 10.

Chapter 3   This chapter provides definitions of the terms and concepts used in Chapters 4 through 10. It also includes a section on the assumptions and conditions applicable to the examples; this section is intended primarily for instructors.

Chapters 4–10   These chapters contain a number of examples representing actual clinical situations. Each chapter presents examples related to a specific joint or joint complex. For the convenience of the reader, the examples in all chapters have been categorized according to their mechanical content in a list following the Contents.

In each of these seven chapters, the problems accompanying the examples are arranged in order of increasing difficulty. The headings in the left margin indicate the mechanical content of each example. All examples begin with a "clinical" description of the problems to be solved to allow the reader to select problems of greatest interest. Suggested topics for discussion follow most examples. As a rule, the discussion questions are designed to investigate:

- What the answers to the problems imply
- How the answers compare with actual experience
- Whether the answers are reasonable

In each chapter, the section entitled Answers, Solutions, and Discussion contains the answers to all the problems in the chapter, as well as more extensive solutions to selected problems. Answers are also given for the discussion questions. Most of the chapters conclude with a commentary, intended mainly for instructors and trained therapists. Reference is made to the values for loads found in the biomechanics literature, and differing research results are compared. This information is presented to illustrate the difficulty in comparing apparently similar measurements and to generate debate on various treatment methods. Certain assumptions made in the solutions to the problems are also discussed.

Three appendices appear at the end of the book.

Appendix A   This appendix contains the body segment parameters used in the examples, together with surface landmarks associated with the anatomic centers and axes of motion of various joints. The readers can use these data to design their own problems.

Appendix B   This appendix lists the units of the International System of Units (SI) used in the examples and gives some British-metric equivalents.

Appendix C   This appendix contains graphs showing the maximum isometric torque on a joint produced by certain muscle groups. The curves of these graphs show how isometric muscle strength varies throughout the range of motion of a joint or joint complex. The graphs can be used in planning exercise programs and discussing the effectiveness of specific exercises.

Introduction

The Subdivisions of Mechanics

Mechanics is the study of motion of substances (solids, liquids, or gases). The links of the locomotor system most closely resemble solid bodies, and thus solid body mechanics can be applied.

A solid body is changed in shape (deformed) when subjected to a load. Mechanics of materials is the aspect of mechanics that studies the relationship between external loads acting on the body and the internal deformations of the body caused by those loads. The extent of deformation of the body depends to various degrees on the shape of the body; the properties of the material composing the body; and the properties of the externally applied load, which can be further analyzed in terms of the point of application, magnitude, and direction of the load and the speed at which this load is applied.

In solving problems in mechanics, the deformations are often so small that they can be disregarded with no noticeable effect on the value of the quantities sought. In such cases, the solid body is considered to be rigid; that is, the distance between the particles (molecules) of the body is considered to remain unchanged when forces are applied to the body. Considering the body as rigid involves a simplification. The use of such an idealization makes dealing with mechanical problems involving the locomotor system simpler than would otherwise be possible. Whether such an idealization is allowable depends on the degree of precision needed for solving the particular problem.

Rigid body mechanics, a branch of classical mechanics, is based on Newton's laws and involves the study of motion of rigid bodies and the interrelationships among the forces acting on these rigid bodies. Rigid body mechanics can be subdivided into kinematics and kinetics. Kinematics describes the motion of a body without reference to the forces causing the motion. Kinetics is concerned with the motion of the body in connection with the forces acting on the body. The area of mechanics that treats bodies in motion is traditionally called dynamics. A special case of dynamics, statics, treats bodies at rest. A body is considered at rest when the change in its speed or direction of motion (acceleration) is zero. Statics deals with the composition and resolution of forces, the calculation of torque, and the setting of conditions for static equilibrium.

Simplifications and Limitations Used in the Examples in Chapters 4 through 10

Figure 1 shows the areas of mechanics into which this collection of examples falls. All examples are limited to rigid body mechanics. The link system of the skeleton is considered to be a system of rigid bodies joined together at the center of the joints, as in the toy figure called a jumping jack. In all examples a static state is assumed; that is, any motion that takes place is linear and at a constant speed.

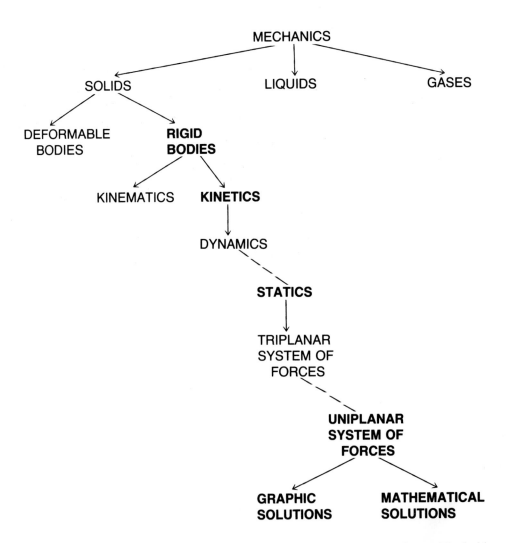

Figure 1. The relationship among various subdivisions of mechanics is shown. The bold-faced terms indicate the areas of mechanics, with their special conditions, that apply to the examples in this book.

In kinetic studies of the locomotor system, it is generally advisable to study forces in all three planes: frontal, sagittal, and transverse. If the magnitudes of the forces acting on the joint are substantially greater in one plane than in the other two, however, forces can be considered to act in this plane only without serious errors in the force calculations. In the problems presented in this book, the calculations are made for forces acting in one plane only.

Problems involving the resolution and composition of forces can be solved graphically by drawing, or mathematically by calculation. Although the emphasis in this book is on the graphic method of solving these problems, the mathematical method is also presented for selected problems.

Recommended Reading

As a supplement to this book, we recommend the following textbook on materials science and the kinematics and kinetics of the skeletal system: Frankel, V.H., and Nordin, M.: *Basic Biomechanics of the Skeletal System*, Lea & Febiger, Philadelphia, 1980. Although a thorough reading of the entire volume is suggested, the reader is directed in particular to Chapter 1, Biomechanics of Whole Bones and Bone Tissue; Chapter 2, Biomechanics of Joint Cartilage; and Chapter 3, Biomechanics of Collagenous Tissues. These chapters cover the loading behavior of the materials composing the skeletal system, a knowledge of which is important in the treatment of musculoskeletal disorders. Chapter 10, Biomechanics of the Lumbar Spine, also includes a discussion of the behavior of the lumbar spine under loading.

Review of Mathematics

1.1. Review Problems

The following review problems represent typical equations used in the problems in Chapters 4 through 10. Working these problems will allow the readers to refresh their mathematical skills and to identify areas of mathematics in which further knowledge and practice are needed. If the readers encounter difficulty in solving problems 1 through 6, they should review a basic algebra textbook. If they have difficulty with problems 7 through 10, they should review the formulas and concepts given in Section 1.2. The readers will need to use either a table of trigonometric functions or a calculator to obtain the value of a sine, cosine, tangent, or cotangent of an angle.

Solve the following problems:

1. $6x = 36$

2. $0.5x + 0.25 \cdot 8 - 15 = 0$

3. $2(x - 3) + 8 = 28$

4. $20x - 2 - 5(2x + 10) = 0$

5. $\dfrac{x}{5} + \dfrac{1}{2} = \dfrac{2}{3}$

6. $x^2 + 21^2 = 37^2$

7. The sides of a right triangle are 4 and 3 cm long. How long is the hypotenuse?

8. In the right triangle ABC in Figure 1.1, side a is 6 cm and side c is 12 cm. How large is angle A?

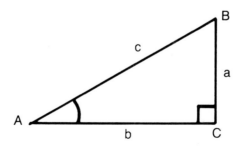

Figure 1.1.

9. In the right triangle ABC in Figure 1.2, the hypotenuse is 10 cm and angle A is 20°. How long are the sides?

10. In the triangle ABC in Figure 1.3, side b=5.0 cm; side c=7.0 cm; and angle A=60°. How long is side a?

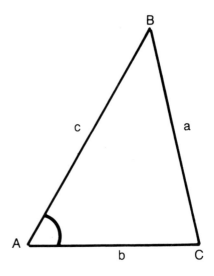

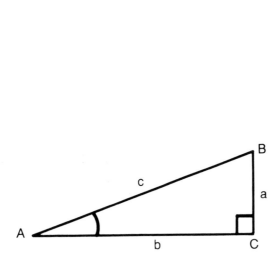

Figure 1.2.

Figure 1.3.

1.2. Useful Formulas and Concepts

Alternate angle. If two parallel lines are intersected by a third line, the alternate angles will be equal (Figure 1.4).

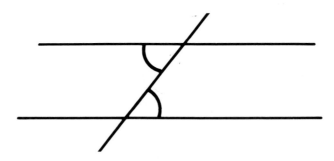

Figure 1.4.

Sum of the angles. The sum of the angles in a triangle is 180°.

Outer angle. The outer angle of a triangle is equal to the sum of the opposite inside angles (Figure 1.5).

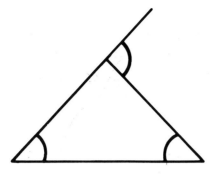

Figure 1.5.

Pythagorean theorem. In a right triangle, the square of the hypotenuse is equal to the sum of the squares of the sides.

$$c^2 = a^2 + b^2 \text{ (see Figure 1.1)}$$

Sine. The sine of an angle in a right triangle is the ratio of the opposite side to the hypotenuse.

$$\sin A = \frac{a}{c} \text{ (see Figure 1.1)}$$

Cosine. The cosine of an angle in a right triangle is the ratio of the adjacent side to the hypotenuse.

$$\cos A = \frac{b}{c} \text{ (see Figure 1.1)}$$

Tangent. The tangent of an angle in a right triangle is the ratio of the opposite side to the adjacent side.

$$\tan A = \frac{a}{b} \text{ (see Figure 1.1)}$$

Cotangent. The cotangent of an angle in a right triangle is the ratio of the adjacent side to the opposite side.

$$\cot A = \frac{b}{a} \text{ (see Figure 1.1)}$$

Supplementary angle. The supplementary angle of an angle is equal to 180° minus the angle. The sine of an angle is equal to the sine of the supplementary angle.

$$\sin A = \sin (180° - A)$$

The cosine of an angle is equal to the negative cosine of the supplementary angle.

$$\cos A = -\cos (180° - A)$$

Cosine theorem. The square of one side of a triangle is equal to the sum of the squares of the other two sides, minus twice the product of these sides and the cosine of the intermediate angle (see Figure 1.3).

$$a^2 = b^2 + c^2 - 2bc \cdot \cos A$$

3

Sine theorem. The sides of a triangle are proportional to the sines of the opposite angles (see Figure 1.3).

$$\frac{a}{\sin A} = \frac{b}{\sin B} = \frac{c}{\sin C}$$

1.3. Answers and Solutions to the Review Problems

Listed below are the correct answers to the problems given in 1.1, followed by the complete solutions for the problems.

1. $x = 6$

2. $x = 26$

3. $x = 13$

4. $x = 5.2$

5. $x = \dfrac{5}{6}$

6. $x = 30.5$

7. 5 cm

8. $30°$

9. 3.4 and 9.4 cm

10. 6.2 cm

Problem 1

$$x = \frac{36}{6} \qquad x = 6$$

Problem 2

$$0.5x + 0.25 \cdot 8 - 15 = 0$$
$$0.5x = 15 - 0.25 \cdot 8$$
$$0.5x = 13$$
$$x = \frac{13}{0.5} \qquad x = 26$$

Problem 3

$$2(x-3) + 8 = 28$$
$$2x - 6 + 8 = 28$$
$$2x = 28 + 6 - 8$$
$$2x = 26$$
$$x = 13$$

Problem 4

$$20x - 2 - 5(2x + 10) = 0$$
$$20x - 2 - 10x - 50 = 0$$
$$10x = 52$$
$$x = 5.2$$

Problem 5

$$\frac{x}{5} + \frac{1}{2} = \frac{2}{3}$$
$$\frac{x}{5} = \frac{2}{3} - \frac{1}{2}$$
$$\frac{x}{5} = \frac{(2 \cdot 2 - 1 \cdot 3)}{6}$$
$$\frac{x}{5} = \frac{(4 - 3)}{6}$$
$$\frac{x}{5} = \frac{1}{6}$$
$$x = \frac{5}{6}$$

Problem 6

$$x^2 + 21^2 = 37^2$$
$$x^2 = 37^2 - 21^2$$
$$x = \sqrt{928}$$
$$x = 30.5$$

Problem 7

Assume that the hypotenuse is c cm long. According to the Pythagorean theorem:

$$c^2 = 4^2 + 3^2$$
$$c^2 = 16 + 9$$
$$c = \sqrt{25}$$
$$c = 5$$

Problem 8

$$\sin A = \frac{a}{c}$$
$$\sin A = \frac{6}{12}$$
$$\sin A = 0.5$$
$$A = 30°$$

Problem 9

$$\sin A = \frac{a}{c}$$

$$\sin 20° = \frac{a}{10}$$

$$a = 10 \cdot \sin 20°$$

$$a = 10 \cdot 0.342$$

$$a = 3.42$$

Problem 10

According to the cosine theorem:

$$a^2 = 5.0^2 + 7.0^2 - 2 \cdot 5.0 \cdot 7.0 \cdot \cos 60°$$

$$a^2 = 25.0 + 49.0 - 35.0$$

$$a^2 = 39.0$$

$$a = \sqrt{39}$$

$$a = 6.2$$

Basic Biomechanical Concepts

2.1. Force

A force is a physical quantity that can accelerate and/or deform a body. The common unit of measurement of force is the newton (N). One newton is the force required to accelerate a mass of 1 kg by 1 m/s². The acceleration of a body produced by the earth's gravity is approximately 9.8 m/s² (varying slightly from place to place on the earth's surface); thus, the gravitational force acting on 1 kg is about 9.8 N. For the sake of simplicity, this value has been rounded off to 10 N in the examples in this book.

A force implies both magnitude and direction and is thus a vector quantity. Examples of other vector quantities are weight, velocity, and acceleration. Quantities that imply only magnitude are scalar quantities. Examples are mass, temperature, length, and time.

A complete analysis of the effect of a force on a body requires knowledge of:

- The point of application of the force.
- The direction of the force (the term direction includes line of application and sense).
- The magnitude of the force.

When all three aspects of a force are known, the force is designated as a vector.

In problems in mechanics a force is commonly depicted graphically by an arrow (Figure 2.1). The base of the arrow represents the point of application of the force, and the orientation of the arrow represents the line of application. The tip of the arrow shows the sense of the force, and the arrow's length represents the magnitude. Extension of the arrow in either direction represents an extension of the line of application. The point of application of a force acting on a rigid body may be displaced along the line of application without changing the effect on the body (the displacement theorem; see Section 2.6, Fundamentals of Rigid Body Mechanics).

In the human body, a force produced by a muscle acts in the longitudinal direction of the tendon of that muscle (Figure 2.2). For example, the force produced by the quadriceps muscle acts on the tibia through

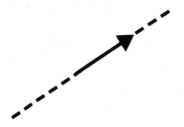

Figure 2.1. Arrow representing a vector. The length of the solid line indicates the magnitude of the vector. The tip of the arrow and the orientation of the line indicate the direction (sense and line of application). The base of the arrow represents the point of application. The broken lines represent extensions of the line of application.

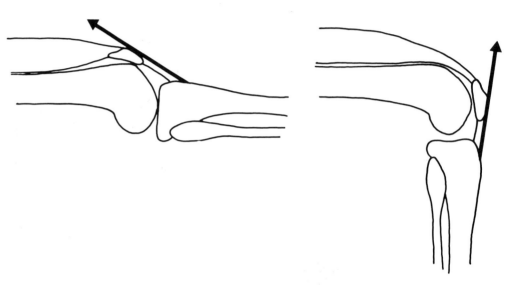

Figure 2.2. The quadriceps muscle force acts on the tibia through the patellar tendon. The muscle force parallels the patellar tendon longitudinally.

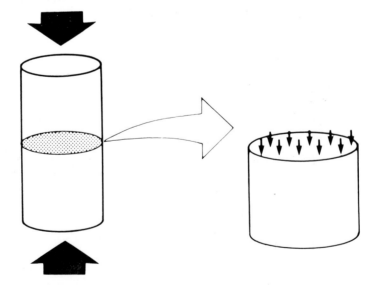

Figure 2.3. When a compressive force (large arrows) is applied to a structure, compressive stress (small arrows) develops within the structure on a plane perpendicular to the applied load. (From Frankel and Nordin, 1980.)

the patellar tendon. The line of application of the force parallels the long axis of the patellar tendon.

Forces directed perpendicularly toward or away from any cross-sectional surface are called normal forces. A force directed toward the surface is called a compressive force (Figure 2.3), and a force directed away from the surface is designated as a tensile force (Figure 2.4). A force acting parallel to the cross-sectional surface is termed a tangential force, or shear force (Figure 2.5).

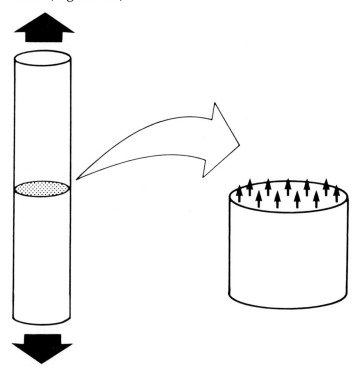

Figure 2.4. When a tensile force (large arrows) is applied to a structure, tensile stress (small arrows) develops within the structure on a plane perpendicular to the applied load. (From Frankel and Nordin, 1980.)

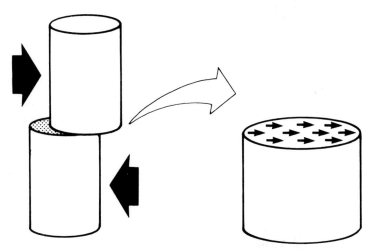

Figure 2.5. When a shear force (large arrows) is applied to a structure, shear stress (small arrows) develops within the structure on a plane parallel to the applied load. (From Frankel and Nordin, 1980.)

9

Distinctions must be made between tensile force and tensile stress, between compressive force and compressive stress, and between shear force and shear stress. The term stress expresses force per unit of area. The basic unit for measuring stress is the pascal (Pa), also known as the newton per meter squared (N/m^2).

Composition of Forces

Composition of forces refers to the combining of forces to determine their net effect. Two or more forces with a common point of application can be replaced by a single force. This replacement force is called the resultant, or resultant force, and the original forces are called components. If the original forces are in equilibrium, all force components cancel each other out, and there is no resultant. The rules for the composition of linear forces (forces acting along the same line of application) and nonlinear forces (forces acting along concurrent lines of application in the same plane) are given below.

Forces Acting along the Same Line of Application

Linear forces acting along the same line of application and with the same sense are added to one another.

Example

A person is exercising the knee extensor muscles with the use of two 5-kg weights suspended from the ankle, each producing a force of 50 N (Figure 2.6A). The two weights can be replaced by one 10-kg weight, producing a force of 100 N (Figure 2.6B). The resultant is obtained graphically by drawing arrows representing the forces to a certain scale, placing the base of one arrow at the tip of the other arrow in a straight line. The length of the line represents the magnitude of the resultant.

Linear forces acting along the same line of application but with an opposite sense are subtracted from each other.

Example

The person in the previous example now receives help in the form of an upward force of 70 N with the same line of application and point of application as the 100-N force produced by the 10-kg weight (Figure 2.7A). Since the forces act with an opposite sense, the resultant is found by subtraction. The magnitude of this resultant force, the "exercise weight," is thus 30 N (Figure 2.7B).

When three or more linear forces act along the same line of application but with opposite senses, all forces acting in one direction are added, and all those acting in the opposite direction are subtracted. The designation of the direction as positive or negative is arbitrary.

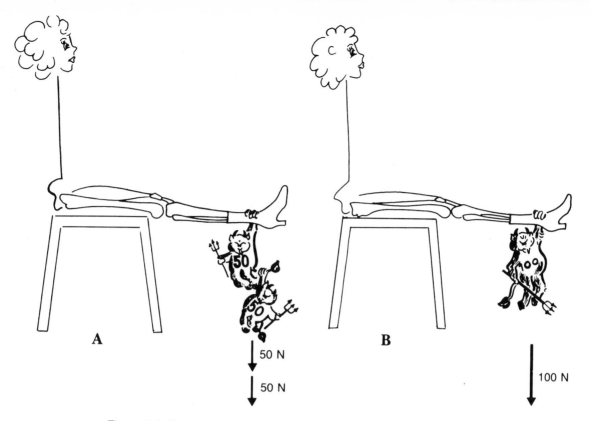

Figure 2.6. Composition of linear forces acting in the same direction. The two 50-N forces are cumulative (Figure A) and can be replaced by a resultant force of 100 N (Figure B).

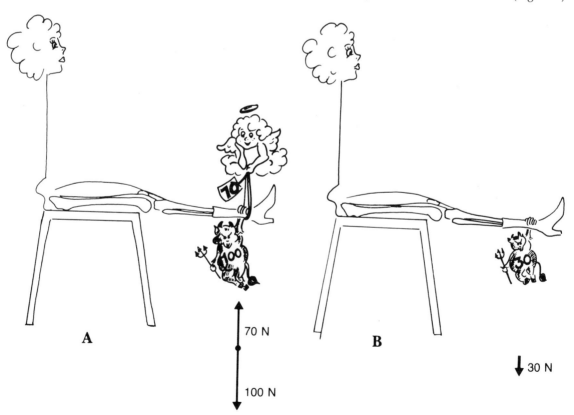

Figure 2.7. Composition of linear forces acting along the same line of application but with an opposite sense. The two forces of 100 N and 70 N are subtracted (Figure A) and can be replaced by a downward resultant force of 30 N (Figure B).

Example

In a tug-of-war, the forces acting on the rope are added and subtracted (Figure 2.8). In this example, forces directed to the left are designated as positive, and those directed to the right are negative. Assume that person A pulls with a force of 900 N, person B with 500 N, person C with 200 N, person D with 400 N, and person E with 700 N. The resultant is obtained by the following calculation:

$$900 + 500 - 200 - 400 - 700 = 100 \text{ N}$$

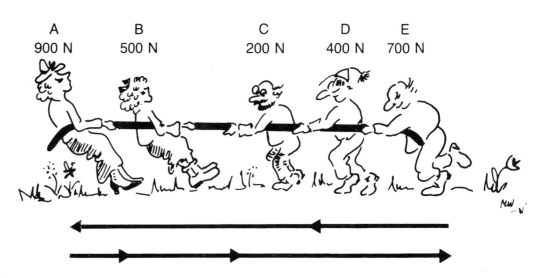

Figure 2.8. Composition of more than two linear forces acting along the same line of application but with opposite senses. All forces directed to the right are subtracted from all forces directed to the left. The five forces acting on the rope can be replaced by a 100-N resultant force acting to the left. (From Williams and Lissner, 1977.)

The resultant has a magnitude of 100 N and is directed to the left; thus, in this example, the women win the tug-of-war.

Forces Acting along Concurrent Lines of Application in One Plane

The magnitude and direction of the resultant (R) of two forces (A and B) with the same point of application can be determined by means of the parallelogram method. In this method, the sides of the parallelogram represent the components, and the diagonal represents the resultant (Figure 2.9). The magnitude of the resultant is determined graphically by drawing the parallelogram of forces to a selected scale and measuring the length of the diagonal.

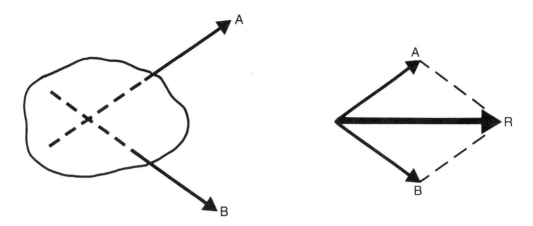

Figure 2.9. Composition of forces by means of the parallelogram method. The resultant (R) is represented by a diagonal of the parallelogram in which the force components (A and B) are represented by the sides.

Example   Two nurses are pushing a bed (Figure 2.10A). Nurse G pushes with a force of 300 N and nurse T, with a force of 400 N. The resultant (R) of these forces is therefore 500 N, which acts at a 37° angle to the force component produced by nurse T. Nurse R can therefore replace the other two nurses (Figure 2.10B).

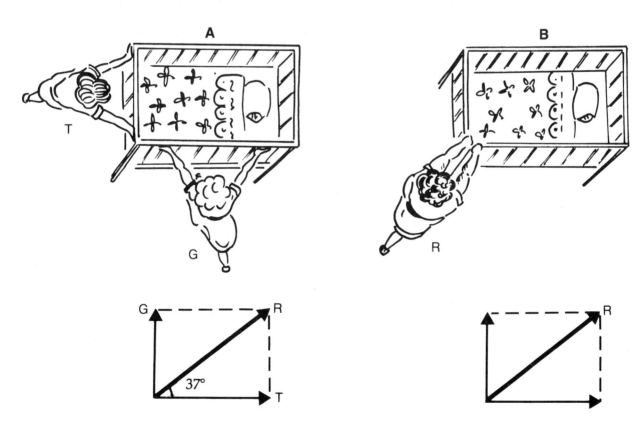

Figure 2.10. Composition of forces by means of the parallelogram method. The force components acting at right angles to one another (G and T) can be replaced by a resultant force (R) directed obliquely to the right (Figure A). Nurse R directs her force more effectively (Figure B). (From Williams and Lissner, 1977.)

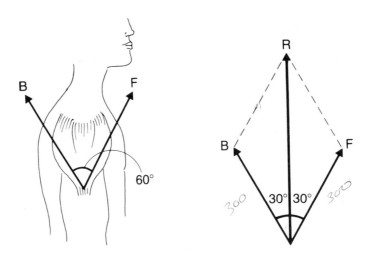

Figure 2.11. Composition of forces by means of the parallelogram method. Two equal force components (F and B) produced by the anterior and posterior portions of the deltoid muscle, which form a 60° angle with one another, can be replaced by a resultant force (R) acting at a 30° angle from each of the two components. (From Williams and Lissner, 1977.)

Example

The anterior and posterior portions of the deltoid muscle each pull with a force of 300 N, and the angle between the force components (F and B) is 60° (Figure 2.11). The resultant is therefore about 520 N and acts in a direction such that the shoulder joint is abducted. The magnitude of the resultant can be found algebraically by means of the Pythagorean theorem or by use of simple trigonometry.

The polygon method is a simplification of the parallelogram method. This method is particularly applicable when more than two forces act on a rigid body. Arrows representing the forces acting on the body are drawn, base to tip, with their lengths proportional to the magnitudes of the forces and their orientation corresponding to the directions of the

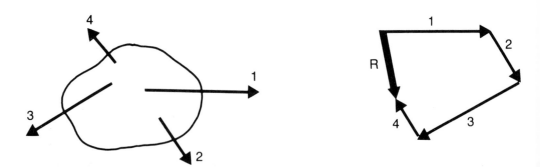

Figure 2.12. Composition of forces by means of the polygon method. The magnitude and direction of the resultant (R) are obtained by connecting the base of the first force component with the tip of the last force component. This "closing side" of the polygon of forces represents the resultant.

forces. The order in which the arrows are drawn is of no consequence. To close the polygon, the tip of the last arrow drawn is connected to the base of the first arrow with a straight line. This "closing side" in the polygon of forces thus formed represents the resultant. The arrow representing the resultant is directed away from the base of the first arrow drawn and toward the tip of the last arrow. The magnitude of the resultant can be scaled from the drawing (Figure 2.12).

Example   Inman (1947) determined the direction of the resultant hip abductor muscle force by determining the lines of application of the individual muscle forces and their proportional magnitudes. The lines of application of the individual muscle forces were determined from an x-ray image on which the point of origin and point of attachment of each muscle were connected with a straight line. The magnitudes of the forces were obtained by weighing each individual muscle and expressing the weight of each as a proportion of the combined weight of all the muscles.

The ratios for the muscle weights were as follows: the tensor fasciae latae muscle, the gluteus minimus muscle, and the gluteus medius muscle, 1:2:4.

With this knowledge of the directions of the individual muscle forces, as well as their relative magnitudes determined from their proportional weights, a polygon of forces could be constructed and the magnitude and direction of the resultant abductor muscle force could be found (Figure 2.13).

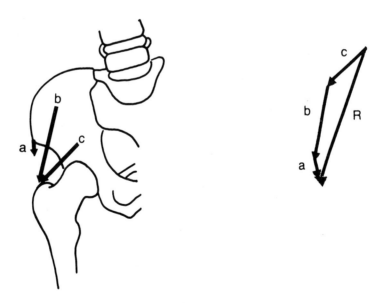

Figure 2.13. Composition of forces by means of the polygon method. When the hip joint is slightly abducted, the most important hip abductor muscles are the tensor fasciae latae muscle (a), the gluteus medius muscle (b), and the gluteus minimus muscle (c). The magnitude and direction of the resultant abductor muscle force (R) are determined by drawing the individual muscle forces from base to tip and then closing the polygon with the resultant force. (From Inman, 1947.)

Resolution of Forces

Resolution refers to splitting a given force into two or more components. An infinite number of components can be found, depending on what is to be studied. The resolution of forces makes use of the parallelogram method. The original force (the force to be resolved) is a diagonal in the parallelogram. Figure 2.14 shows two possible ways of resolving a given force. It is practical to resolve the forces in a system of rectangular coordinates. In a plane force system, vertically directed forces (F_y) and horizontally directed forces (F_x) can then be obtained.

 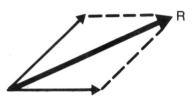

Figure 2.14. Resolution of forces by means of the parallelogram method. Two arbitrarily chosen ways of resolving force R into components are shown.

Example

As a person's heel strikes the ground during walking, a force is directed obliquely downward toward the ground (Figure 2.15). This force can be resolved into a vertical force component (F_y) and a horizontal force component (F_x). The relative magnitudes of these forces depend on the length of the step taken. The longer the step, the greater is the F_x component. For the heel to be prevented from slipping forward, the horizontal force

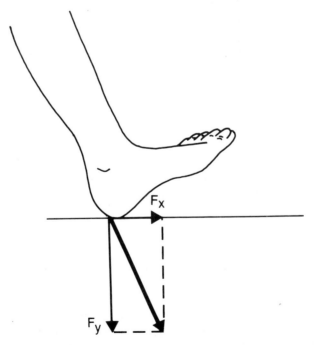

Figure 2.15. Resolution of forces by means of the parallelogram method. The force directed obliquely downward from the heel can be resolved into a vertically directed force component (F_y) and a horizontally directed force component (F_x).

component must be counteracted by the friction force between the ground and the foot.

In situations involving a force directed toward a surface, e.g., a joint surface, the force is often resolved into a component perpendicular to the surface (a compressive or tensile force) and a component tangential to the surface (a shear force). If no friction exists, no shear forces are produced between the joint surfaces. In this case, any shear forces are counteracted by the surrounding soft tissues.

Example

A person stands with normal posture (Figure 2.16A) and then with a pronounced lordosis (Figure 2.16B). The weight of the body above the sacrum (W) can be resolved into a compressive force component perpendicular to the sacral joint surface (C) and a shear force component parallel to the sacral joint surface (S). The shear force (S) produced by the weight of the upper body (W) against the sacrum increases when the curvature of the spine becomes more pronounced (i.e., when the angle of inclination of the sacrum to the horizontal plane increases).

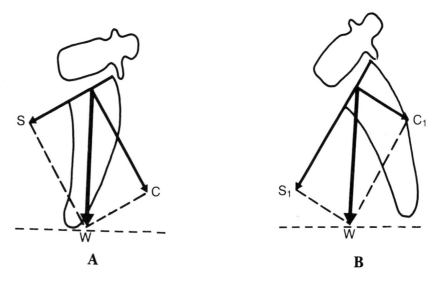

A   **B**

Figure 2.16. The body weight above the sacrum (W) produces a shear force (S) and a compressive force (C) on the sacrum. The angle of inclination of the sacrum to the horizontal plane is smaller in Figure A than in Figure B; thus, S is smaller than S_1 and C is larger than C_1. (From Williams and Lissner, 1977.)

Force Equilibrium (Translatory Equilibrium): First Equilibrium Condition

If the resultant of a number of forces acting on a rigid body is zero, the forces are in equilibrium according to the first condition of equilibrium, called force or translatory equilibrium. When force equilibrium is achieved, the rigid body is not displaced. Force equilibrium means that the sum of all forces is zero ($\Sigma F = 0$). If the forces have been resolved into vertically and horizontally directed force components, then:

- All upward forces minus all downward forces equals zero.

  $\Sigma F_y = 0$

- All forces directed to the right minus all forces directed to the left equals zero.

  $\Sigma F_x = 0$

Figure 2.17. Force equilibrium. In this example, the forces directed downward are equal to the upward-directed force; i.e., the sum of the forces is zero. (The weight of the board is disregarded.)

When a polygon of forces is constructed to depict the forces acting on a body in equilibrium, the tip of the last arrow drawn connects with the base of the first arrow. Thus, the polygon closes without a resultant, demonstrating that the first equilibrium condition has been fulfilled.

Example

For force equilibrium to be achieved in the case of a seesaw, the force directed upward toward the board must equal the sum of the forces directed downward (Figure 2.17). (Note that achievement of the first equilibrium condition is not enough for the board to balance, a concept that is explained in Section 2.2, Torque.)

Example

A man is standing on one leg, and the pelvis is balanced around the hip joint by the weight of the upper body and the abductor muscle force (Figure 2.18). The weight of the upper body (T) on one side of the hip and the abductor muscle force (A) on the other side compress the acetabulum against the femoral head. The femoral head responds with an equal but opposite force (R), the hip joint reaction force. The magnitude of R is obtained by means of the polygon method.

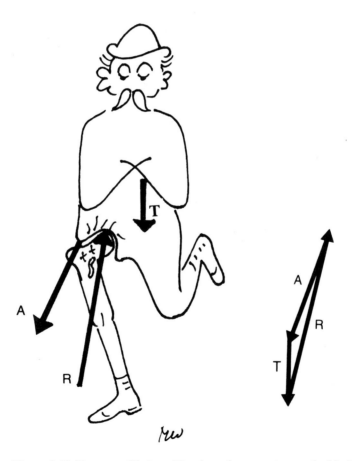

Figure 2.18. Force equilibrium. The three forces acting on the hip joint balance one another to keep the body in equilibrium. These forces are the abductor muscle force (A), the weight of the upper body (T), and the hip joint reaction force (R). The polygon of forces forms a closed figure with no resultant. (From Frankel and Nordin, 1980.)

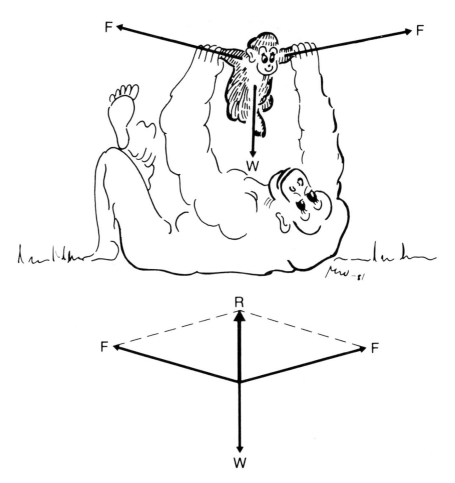

Figure 2.19. Force equilibrium. The baby monkey's weight (W) is equal in magnitude and opposite in sense to the resultant (R) of the two forces (F) exerted on the baby monkey's arms.

Example

A baby monkey is lifted by its arms (Figure 2.19). With the monkey in equilibrium, the resultant (R) of the two forces (F) exerted on the arms is equal in magnitude and opposite in sense to the weight of the monkey (W).

2.2. Torque

Even if force equilibrium is achieved (i.e., the sum of the forces acting on a body equals zero) and the first condition of equilibrium is met, the body may not be in a state of complete equilibrium. Although the body will not displace, the action of certain forces can cause the body to rotate around an axis perpendicular to the plane of the forces; i.e., these forces exert a torque, or moment, on the body.

Example

Two forces equal in magnitude, opposite in sense, and with parallel lines of application form a force couple acting on a body. The resultant of the forces is zero, meaning that the body is not displaced (i.e., the body is in translatory equilibrium). However, the force couple causes the

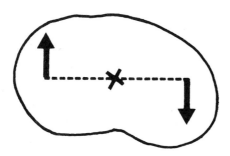

Figure 2.20. Two forces (a force couple) exert a torque (moment) on a body. The broken lines represent the moment arms of the forces relative to an arbitrarily selected center of rotation in the body (marked with an X).

body to rotate around an axis perpendicular to the plane of the forces; that is, together the two forces exert a torque (moment) on the body (Figure 2.20).

The magnitude of the torque is the product of the force and the moment arm (also called the lever arm) (see Fig. 2–20). If the force is expressed in newtons (N) and the moment arm is expressed in meters (m), the unit of measurement of torque is the newton meter (Nm).

The moment arm is defined as the perpendicular distance from the line of application of the force to the center of rotation in the body. The center of rotation is an arbitrarily selected point around which the torque, or moment, acts. A force whose line of application passes through the center of rotation exerts no torque around the center of rotation because there is no moment arm.

In mechanical problems, the center of rotation may be placed any-where, provided that all forces are examined with reference to this point. In biomechanical problems involving forces acting on a joint, the center of rotation is often placed at the anatomic center of motion of the joint. In actuality, this anatomic center of motion shifts during joint motion, and thus the center of rotation is placed at a point somewhere within the region wherein this shift normally occurs.

A torque acts in a certain direction. In a plane system of forces, this direction can be clockwise or counterclockwise. In mechanical problems, the direction is arbitrarily designated as positive or negative.

The magnitude of a torque can be changed in two ways:

- The magnitude of the force can be changed.
- The length of the moment arm can be changed (by shifting the point of application of the force relative to the center of rotation and/or by changing the direction of the force).

Example

A woman carries her purse in her hand at a distance p from the elbow joint (Figure 2.21A). The weight of the purse (P) exerts an extending moment (P·p) on the elbow joint. The point of application of the force is changed when the woman moves the purse to a distance of $\frac{p}{3}$ from

21

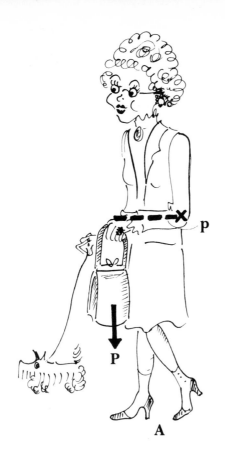

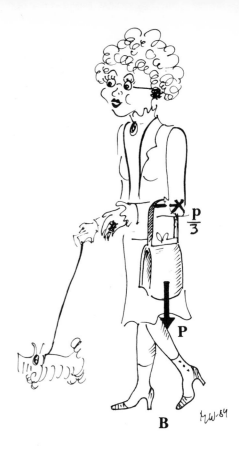

A

B

Figure 2.21. The extending moment on the elbow joint is the product of the force produced by the weight of the purse (P) and its moment arm (p) (the perpendicular distance from the force to the center of rotation of the elbow joint, marked with an X). Thus, the extending moment in Figure A (P · p) is 3 times greater than that in Figure B (P · $\frac{p}{3}$).

elbow joint (Figure 2.21B). This shift reduces the extending moment produced by the weight of the purse by two-thirds. Thus, it becomes easier for this elderly woman to carry her purse.

Moment Equilibrium (Rotatory Equilibrium): Second Equilibrium Condition

Moment equilibrium, or equilibrium according to the second equilibrium condition, is achieved when the sum of the moments tending to rotate the body in one direction minus the sum of the moments tending to rotate the body in the opposite direction equals zero; that is, $\Sigma M = 0$ (Figure 2.22). Thus, during moment equilibrium (also called rotatory equilibrium), no rotation of the body takes place.

Example

A man is standing on one leg, and the hip abductor muscle force counteracts the rotatory effect of the weight of the upper body on the hip joint. The weight of the upper body (T) acting on the medial side of the hip joint rotates the upper body in a clockwise direction. The abductor muscles exert a force (A) on the pelvis on the lateral side of the hip joint and rotate the upper body in a counterclockwise direction. With the body in moment, or rotatory, equilibrium, these two moments balance one another (Figure 2.23).

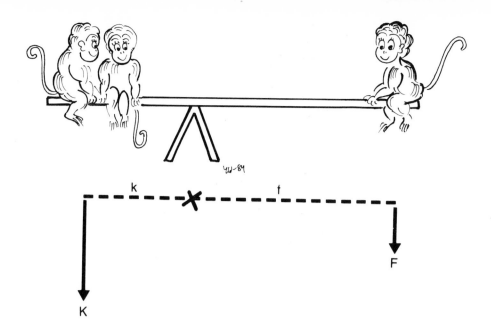

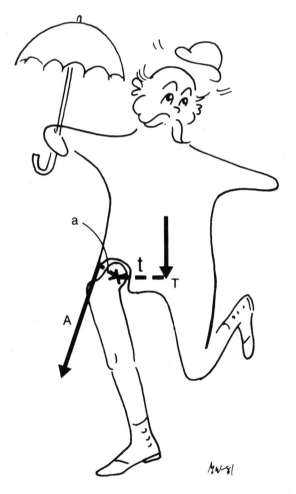

Figure 2.22. Moment equilibrium. With the seesaw in rotatory equilibrium, force K times its moment arm k is equal to force F times its moment arm f. (Kk − Ff = 0, or Kk = Ff.)

Figure 2.23. Moment equilibrium. The weight of the body (T) above the hip joint of the standing leg times its moment arm (t) is balanced by the abductor muscle force (A) times its moment arm (a). The center of rotation of the hip joint is marked with an X.

External and Internal Torque

External torque is the torque produced by the weight of a body segment times its moment arm, or by an externally applied load (such as manual resistance, crutches or a cane, a weight attached to the body, or an object lifted or carried) times its moment arm.

Internal torque is the torque produced by forces associated with internal structures (e.g., forces exerted by muscles or transmitted through tendons and other soft tissues) times their moment arms (Brodin and Moritz, 1971). In Figure 2.23, the product of the weight of the upper body and its moment arm (Tt) constitutes the external torque, whereas the product of the abductor muscle force and its moment arm (Aa) constitutes the internal torque.

The magnitude of the internal torque produced by a muscle force depends on:

- The magnitude of the contraction force of the muscle, which is influenced to a large extent by the length of the muscle relative to its length at rest (the length-tension relationship) and by the number of active motor units in the muscle.
- The length of the moment arm of the muscle force, which varies among muscles and for the same muscle at different joint angles (Figure 2.24).

A muscle that has a small physiologic cross-sectional area but is located at a distance from the center of motion of a joint can exert as large a moment on the joint as can a muscle with a large cross-sectional area but a less advantageous location (nearer the center of motion).

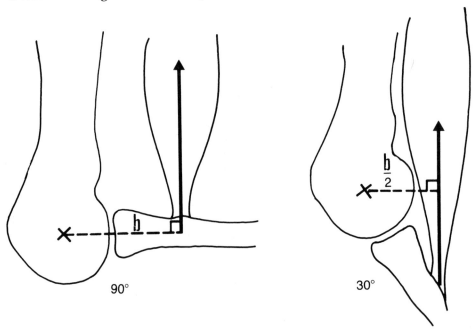

90°   30°

Figure 2.24. The moment arm (b) of the biceps muscle force at 90° and 30° of elbow flexion. The length of the moment arm is influenced by the angle of the elbow joint and in this example is half as long at 30° as at 90° of elbow flexion.

this angle change.
joint angle & the langueur or
tension of produced by
this is the force or
flexing & extension
of the muscle

The maximum internal torque produced by isometric muscle exercise varies with the joint angle because of changes in the length-tension relationship of the muscle and changes in the length of the moment arm of the muscle force. The resulting internal torque can be plotted graphically as a function of the joint angle. The curves obtained show how the internal torque varies throughout the range of joint motion. Appendix C contains graphs of the maximum isometric torque for a number of muscle groups.

2.3. Center of Gravity

The center of gravity of a body is the point at which total mass of the body is considered concentrated. The body is balanced in every direction around this point. The concept of a center of gravity involves a great simplification in mechanical calculations. All of the numerous small forces produced by the action of gravity on the molecules of the body are replaced by a single force applied at the center of gravity of the body and equal to the weight of the body. In symmetrical bodies, such as a ball or a circular disc, the center of gravity is always at the geometric, or symmetrical, center. When the human body is in the upright position,

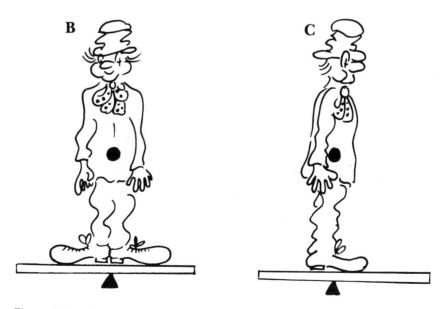

Figure 2.25. The location of the center of gravity (solid dot) in the craniocaudal direction (Figure A), mediolateral direction (Figure B), and ventrodorsal direction (Figure C).

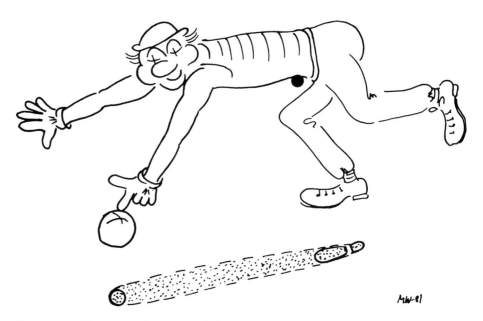

Figure 2.26. The center of gravity (solid dot) need not fall within the body mass but must fall within the region of support (shaded area) if balance is to be maintained.

its center of gravity lies just anterior to the first sacral vertebra (Carlsöö, 1972). Figure 2.25 shows the center of gravity for the human body in the craniocaudal, ventrodorsal, and mediolateral directions.

For balance to be maintained in a body, the center of gravity must fall within the region of support. It need not lie inside the body mass, however (Figure 2.26).

Biomechanical calculations often require a knowledge of the location of the center of gravity within individual body segments. In the examples in this book, data from Dempster (1955) have been used. (See Appendix A.)

2.4. Free Body Diagram

A useful method for calculating forces and moments acting on a body, or a portion of a body, involves the construction of a free body diagram. The body segment of interest is considered in isolation from the rest of the body, from the supporting surface, and from other points of contact. The omitted parts of the body are replaced by the force or forces produced by the action of these parts on the body segment under consideration.

The forces acting on the free body are designated on the free body diagram. Before the forces can be designated accurately, the conditions governing the actions of the forces must be set up (for example, equilibrium conditions). The assumptions and idealizations used in the problem must also be determined.

Example

A man is standing on one leg. The forces acting on the hip joint of the standing leg in the frontal plane are to be calculated with the use of a free body diagram. The free body can consist of either the upper body and the lifted leg, or just the standing leg (Figure 2.27).

Figure 2.27. Free body diagrams showing the forces acting on the hip joint in the frontal plane during a one-leg stance. Either the standing leg alone (bottom left) or the upper body and lifted leg (bottom right) can be chosen as a free body. The forces acting on the standing leg are the ground reaction force (W), abductor muscle force (A), the joint reaction force from the acetabulum (R), and the weight of the standing leg (B). The forces acting on the body and lifted leg are the abductor muscle force (A), the joint reaction force from the femoral head (R), and the weight of the upper body (T). (From Frankel and Nordin, 1980.)

2.5. Friction

When an object slides along a surface, its motion is always counteracted by friction. Friction is greatest at the moment that the object is set in motion (full friction) and then decreases somewhat as motion continues (friction of motion). Friction is proportional to the normal force (the force component perpendicular to the surface against which the object slides).

$$F = \mu \cdot N,$$

where F is the friction force, μ is the coefficient of friction, and N is the normal force.

The following are examples of some coefficients of friction (Williams and Lissner, 1977; White and Panjabi, 1978):

- Wood on wood (dry), 0.23–0.50
- Metal on metal (dry), 0.15–0.20
- Metal on metal (lubricated), 0.03–0.05
- Rubber crutch tip on rough wood, 0.70–0.75
- Hard rubber cane tip on clean tile, 0.18–0.22
- Cartilage on cartilage in an animal ankle joint, 0.005

The coefficient of friction often varies between 0 and 1, with 0 indicating the absence of friction. Friction is independent of both speed and the size of the contact surface.

2.6. Fundamentals of Rigid Body Mechanics

Newton's Laws of Motion

The entire field of rigid body mechanics is based on Newton's three fundamental laws of motion.

Newton's First Law of Motion: All Matter Has Inertia

A body at rest remains at rest, and a body in motion remains in uniform linear motion, unless acted upon by an outside force.

Newton's Second Law of Motion: Force Equals Mass Times Acceleration

The acceleration of a body is directly proportional to the force applied to it, and inversely proportional to its mass. The relationship between force and acceleration is given by the following equation:

$$F = m \cdot a,$$

where F is the force acting on the body (N), m is the mass of the body (kg), and a is the acceleration of the body (m/s^2).

Newton's Third Law of Motion: For Every Action There Is an Equal and Opposite Reaction

When a body exerts a force on a second body, the second body exerts an equal but opposite force on the first body.

Useful Theorems in Rigid Body Mechanics

The basic laws of mechanics include not only Newton's laws but also the displacement theorem and the parallelogram theorem.

The Displacement Theorem

The point of application of a force acting on a rigid body can be displaced along the line of application of the force without changing the effect on the body.

The Parallelogram Theorem

The magnitude and direction of the resultant R of two component forces A and B acting with the same point of application are determined by the length and orientation of the diagonal of a parallelogram constructed with the forces as its sides.

Introduction to the Examples in Chapters 4 through 10

3.1. Definition of Terms Used in the Examples

Body segment

A body segment is the body mass located between two centers of motion. For example, a thigh segment extends from the center of rotation of the hip joint to the center of motion of the knee joint and thus is somewhat shorter than the femur.

Center of motion

The center of motion of a joint is the point around which motion of the joint is described. In the examples in this book, the center of motion is placed within the area wherein the anatomic center of motion shifts during joint motion. Surface landmarks used in placing the anatomic centers and the axes of motion of various joints are described in Appendix A.

Joint reaction force

The joint reaction force is the counterresultant of the forces acting on a joint. The counterresultant force acts with the same magnitude as the resultant but in the opposite direction. The joint reaction force passes through the center of motion of the joint and can be resolved into components. One component can be perpendicular to the joint surface (compressive or tensile force), and the other can be parallel to the joint surface (shear force). This resolution requires precise knowledge of the position of the joint surface relative to the joint reaction force.

Link

A link of the skeletal system is the skeletal portion of a body segment. For example, the femur is the link associated with the thigh segment and is somewhat longer than that body segment.

Mass

The mass of a body is the amount of matter that the body contains and that causes it to have weight in a gravitational field. Mass is a scalar quantity, having magnitude but not direction. It is commonly expressed in kilograms (kg).

Muscle strength

Muscle strength is the force developed by a muscle or muscle group to counteract a torque. In practice, the magnitude of the torque, rather than of the muscle force, is measured.

Weight

The weight of a body is the gravitational force exerted on the body, equal to the product of the body's mass and the local value of gravitational acceleration. Weight differs from mass in that it is a vector quantity, having both magnitude and direction. It is commonly expressed in newtons (N).

3.2. Assumptions and Conditions in the Examples

- The links of the skeletal system are considered as rigid bodies. Thus, no deformation takes place to change the distribution of forces.

- Within each problem the forces acting on the joint are considered in one plane only. This simplification is permissible when the magnitudes of the forces in one plane are substantially greater than in the other two planes.

- The joints are assumed to be free of friction (Andersson and Schultz, 1979). The coefficient of friction in a normal joint is 0.005 (White and Panjabi, 1978) and can thus be disregarded.

- The external torque is assumed to be entirely absorbed by the muscle or muscle group designated in each example. A muscle group is considered to exert a force in one direction only. Consequently, forces produced by other muscles, or by ligaments, capsules, and other soft tissues, are disregarded.

- There is assumed to be no antagonistic muscle activity.

- The weights of the body segments as a percentage of the total body weight, as well as the placement of the center of gravity for the body segments, were obtained from the literature.

- The lengths of the moment arms, the directions of the muscle forces, and the orientation of the joint surfaces relative to the designated plane were obtained from the literature when possible or were determined from x-ray films.

- Unless otherwise stated, the moment arm lengths for the external torques were determined from photographs taken for that purpose. The photographs were taken perpendicular to the axis of motion under consideration. To permit calculations, a scale placed at the same distance from the camera as the body part itself was also photographed. The photographs were taken at a sufficient distance from the body part to allow errors in measurement resulting from the depth effect to be disregarded. Subsequent enlargement of the photographs to a suitable size also reduced measurement errors.

The Ankle

The ankle joint is situated between the talus, the tibia, and the fibula. The joint is quite stable because the talus is surrounded by the distal ends of the tibia and fibula. Motion in the ankle joint takes place primarily in the sagittal plane. The transverse axis of rotation passes through the talus (Sammarco et al., 1973).

4.1. Simple Heel Raise. Functional Strength Test of the Plantar Flexor Muscles

Torque

In this example, which illustrates the relationship between external and internal torque, the magnitude of the force transmitted through the Achilles tendon when the heel is raised from the floor is compared with that required for standing on the toes.

Problems

4.1A. A patient weighing 70 kg performs a simple heel raise. How large a force (F) is transmitted through the Achilles tendon when the patient lifts the heel from the floor? The weight of the body is assumed to pass through the heads of the metatarsals of the standing foot, and its moment arm is 15.0 cm. The moment arm of the force through the Achilles tendon is 5.0 cm. Disregard the weight of the foot (Figure 4.1A).

4.1B. How large a force (F) is transmitted through the Achilles tendon when the same patient stands on the toes of one foot? The moment arm of the body weight is 10.0 cm, and the moment arm of the force through the Achilles tendon is 4.5 cm (Figure 4.1B).

Discussion Questions

1. Why is the arrow representing the body weight pointing upward in Figures 4.1A and B?

2. Which requires the larger muscle force: lifting the heel from the floor or standing on the toes?

3. Develop the example further and discuss the phase of the gait cycle in which the force produced by the plantar flexor muscles is the largest.

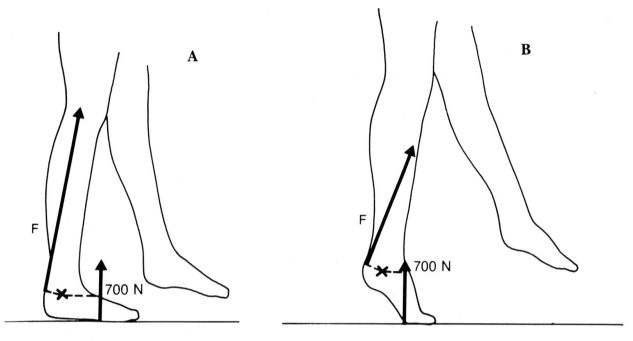

Figure 4.1. The patient performs a simple heel raise. In Figure A, the patient has raised the heel from the floor. In Figure B, the patient is standing on the toes of one foot. The force (F) transmitted through the Achilles tendon balances the body weight (700 N). The center of motion of the ankle joint is marked with an X.

4.2. Answers, Solutions, and Discussion

Answers to the Problems

4.1A. F = 2100 N

4.1B. F = 1556 N

Solution to Problem 4.1A

Find the Achilles tendon force (F) when the patient lifts the heel from the floor. In this example, the counterclockwise moment is positive.

$\Sigma M = 0$ gives the magnitude of the Achilles tendon force F.

$700 \cdot 0.15 - F \cdot 0.05 = 0$

$$F = \frac{700 \cdot 0.15}{0.05}$$

$F = 2100$ N

Answers to Discussion Questions

1. In Figures 4.1A and B, the arrow representing the body weight points upward because the forces being considered in this example are those acting on the foot. The force of the floor against the foot (the ground

reaction force) equals the force of the foot against the floor produced by the body weight, but it acts in an opposite direction.

2. The moment arm of the body weight is longer at the beginning of the heel raise than at the end, and thus the torque is larger. The muscle force required to balance the body weight is therefore larger when one lifts the heel from the floor than when one rises onto the toes.

It has been found experimentally that some patients with reduced strength in the plantar flexor muscles can raise the heel from the floor but cannot rise all the way up onto the toes. An explanation is that the length-tension relationship of the muscle is such that a larger force can be exerted at the beginning of the heel raise. (See External and Internal Torque in Section 2.2 of Chapter 2 and the discussion of examples 8.2 and 8.3 in Chapter 8.) Since the maximum internal torque is dependent on the joint angle, the muscle strength must be tested over the entire range of motion, and not just at the joint angle at which the external torque is maximum.

3. The primary function of the plantar flexor muscles during normal gait is to resist the ventral rotation of the tibia in the stance phase. The largest muscle force is required at the end of the stance phase, since the torque on the ankle joint during dorsiflexion, produced by the body weight, is largest at that time (Stauffer et al., 1977).

4.3. Commentary

Torque on the Ankle Joint during Dorsiflexion

When one stands in a relaxed position, the line of gravity of the body passes a few centimeters anterior to the transverse axis of rotation of the ankle joint. This means that the body weight produces a dorsiflexing torque on the ankle joint. The magnitude of the torque varies between 3 and 24 Nm as a result of body oscillations (Smith, 1957). Therefore, standing with the body weight distributed evenly on both feet requires a certain strength in the plantar flexor muscles.

The dorsiflexing torque during walking is normally about 100 Nm. If this large a torque cannot be achieved, the walking function can still be maintained through a change in the gait pattern. For example, Murray et al. (1978) described the gait pattern of a patient from whom the triceps surae muscle in one leg had been surgically removed. During slow walking, the changes in the gait pattern were very small; but as the walking speed increased, changes became obvious. When the patient walked longer distances and used the stairs, the quadriceps muscle in the surgically treated leg became fatigued. The experience with this patient supports Sutherland's theory (1966) that the plantar flexor muscles indirectly stabilize the knee joint during walking by regulating the dorsiflexion and plantar flexion of the ankle. A reduced muscle force, with consequent changes in gait pattern, changes the load distribution in the joint. Such a change may predispose the joint to arthrosis.

The magnitude of the torque during stair climbing and descending is somewhat higher than that during walking on level ground. The average maximum torque during stair climbing and descending was calculated to be 137 Nm for 10 healthy 28-year-old men weighing an average of 71 kg (Andriacchi et al., 1980).

The Ankle Joint Reaction Force

As a result of contraction of the plantar flexor muscles, the magnitude of the ankle joint reaction force during symmetrical standing closely corresponds to the body weight. During walking, the joint reaction force is largest at the end of the stance phase, and during push off it can reach five times body weight (Stauffer et al., 1977; Seireg and Arvikar, 1975). The ankle joint has a relatively large surface area (11 to 13 cm²) over which this force is distributed (Ramsey and Hamilton, 1976).

Assumptions Made in Example 4.1

1. The Achilles tendon is assumed to absorb the dorsiflexing torque on the ankle joint produced by the body weight. The flexor hallucis longus muscle, the flexor digitorum longus muscle, the tibialis posterior muscle, the peroneus longus muscle, and the peroneus brevis muscle also act as plantar flexors. On the basis of the cross-sectional area of the muscles and length of the moment arms of the muscle forces, Murray et al. (1976) determined that the triceps surae muscle is responsible for 80% of the force produced by all the plantar flexor muscles.

2. The moment arm of the force transmitted through the Achilles tendon is assumed to be 5.0 cm when the foot is in a neutral position and 4.5 cm with the foot in a plantar-flexed position. Murray et al. (1976) cited a moment arm length of 4.8 cm when the foot is in the neutral position. Two different lengths for the moment arms in the problem are cited to illustrate that the moment arm of the force through the Achilles tendon changes when the joint angle changes. The moment arm is somewhat shorter with the foot in a plantar-flexed position than with the foot in a neutral position (Williams and Lissner, 1977).

5

The Elbow

The elbow is composed of three joints: the humeroulnar, the humero-radial, and the proximal radioulnar. Of these joints, the humeroulnar joint contributes the most to motion in the sagittal plane; in this plane the axis of rotation passes through the trochlea (Chao and Morrey, 1978; Youm and Dryer, 1979). In examples 5.1 and 5.2 motion of the humero-ulnar joint is analyzed in the sagittal plane.

5.1. Strengthening the Elbow Extensor Muscles with Dumbbells

Torque

This example can be used in reviewing and discussing functional muscle strength tests and strengthening exercises.

A patient is strengthening the elbow extensor muscles with dumbbells. The exercises are performed in two positions. The external torque produced by the weight of the dumbbell is calculated at different angles of the elbow joint and compared with the maximum internal torque at these joint angles. The maximum internal torque produced by the elbow extensor muscles is presented in the form of a curve (Figure 5.1), obtained during a dynamic strength test (of isokinetic strength) in which the speed of motion was held constant at 90° per second. The test apparatus used was a Cybex II dynamometer.

Problems

5.1A. A patient is lying on his back and holding a 5-kg dumbbell in one hand. The shoulder is flexed 90° (Figure 5.1A). How large is the torque (M) on the elbow joint produced by the weight of the dumb-bell at 120°, 90°, 30°, and 0° of elbow flexion? Each position during extension of the elbow at a low constant speed can be considered to be static. The distance from the center of gravity of the dumbbell to the center of motion in the elbow joint is 0.3 m. Plot the static values for the torque at the four joint angles on the graph shown in Figure 5.1 and compare these values with the curve for maximum isokinetic torque in the figure.

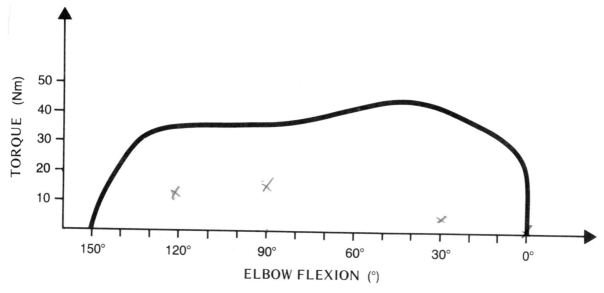

Figure 5.1. Maximum isokinetic torque produced by the elbow extensor muscles during elbow joint motion from full flexion to full extension in one healthy subject. The speed of motion was 90° per second. The horizontal axis shows the joint angle, with 0° representing full extension. The vertical axis shows the maximum torque in newton meters. (From Swedish National Board of Safety and Health, 1980.)

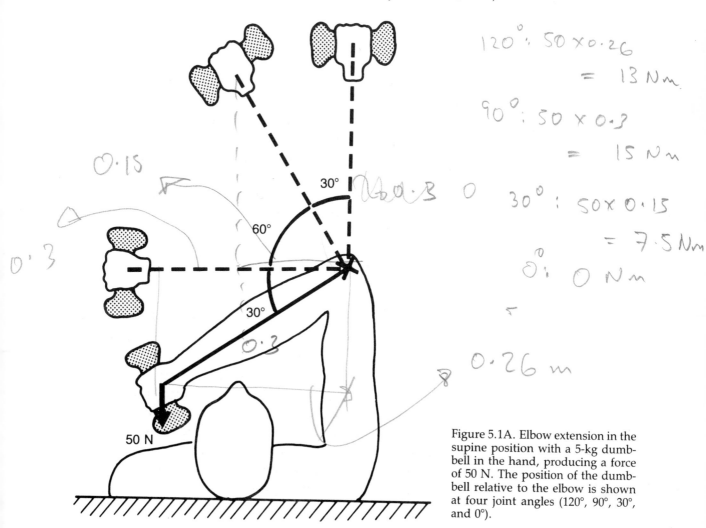

Figure 5.1A. Elbow extension in the supine position with a 5-kg dumbbell in the hand, producing a force of 50 N. The position of the dumbbell relative to the elbow is shown at four joint angles (120°, 90°, 30°, and 0°).

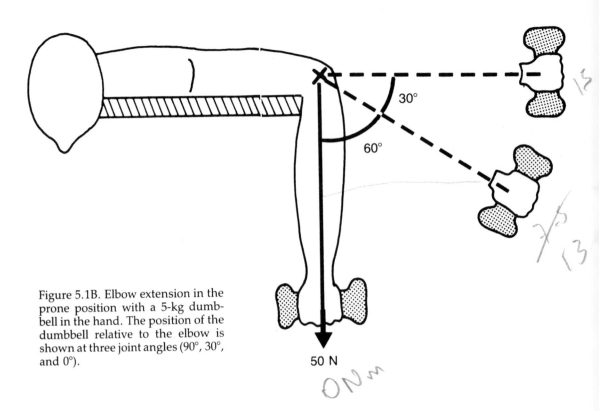

Figure 5.1B. Elbow extension in the prone position with a 5-kg dumbbell in the hand. The position of the dumbbell relative to the elbow is shown at three joint angles (90°, 30°, and 0°).

5.1B. The same patient is now prone with the shoulder abducted 90°. The upper arm is resting against the table, and the forearm is hanging alongside the table (Figure 5.1B). How large is the torque (M) on the elbow joint produced by the weight of the dumbbell at 90°, 30°, and 0° of elbow flexion? The distance from the center of gravity of the dumbbell to the center of motion of the elbow joint is 0.3 m. Plot the static values for the torque at the three joint angles on the graph in Figure 5.1 and compare these values with the curve for maximum isokinetic torque.

Discussion Questions

1. Why, in these two problems, does the magnitude of the external torque change as the angle of the elbow joint changes?

2. Cite some advantages in varying one's position when exercising with weights.

5.2. Performing Push-ups in the Sitting and Prone Positions

Torque

This example gives practice in identifying a free body, determining the direction of the external force acting on the free body, and calculating the torque.

The external torque on the elbow joint during a sitting push-up is compared with the torque during a push-up performed in the prone position.

38

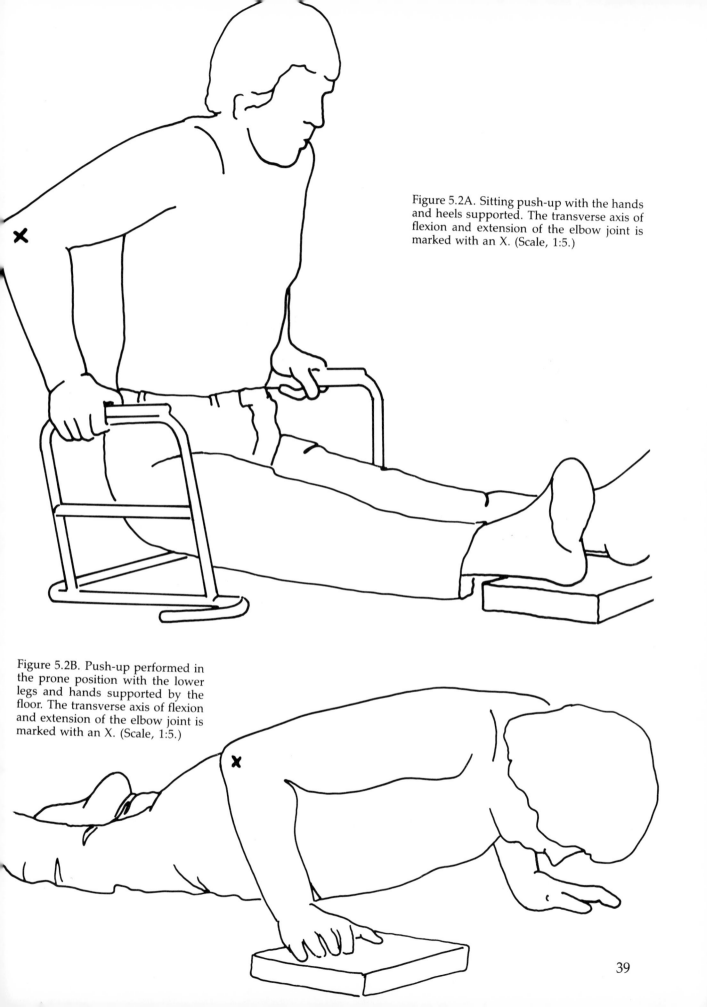

Figure 5.2A. Sitting push-up with the hands and heels supported. The transverse axis of flexion and extension of the elbow joint is marked with an X. (Scale, 1:5.)

Figure 5.2B. Push-up performed in the prone position with the lower legs and hands supported by the floor. The transverse axis of flexion and extension of the elbow joint is marked with an X. (Scale, 1:5.)

39

Problems

5.2A. A paraplegic patient weighing 70 kg is sitting on the floor and is exercising the elbow and shoulder muscles with the aid of a push-up chair (Figure 5.2A). The portion of the body weight transmitted through the heels, determined by means of a force plate, is 100 N acting vertically. The patient exerts a vertical force on the hand-grips, equally distributed between the two hands. Only the heels and hands are supported. How large is the torque (M) on the elbow joint if the moment arm of the force relative to the point of grip in the hand is 0.05 m? Disregard the weight of the forearm. On Figure 5.2A, draw in the external force acting on the right hand. The figure has been drawn so that the length of the moment arm is proportional to the actual length given in the text, on a scale of 1 to 5.

5.2B. The same patient is now performing a push-up in the prone position (Figure 5.2B). The lower legs and hands are supported by the floor. The force plate, placed under the right hand, shows a vertical force of 200 N. The moment arm of this force relative to the center of motion of the elbow joint is 0.08 m. How large is the torque (M)? On Figure 5.2B, draw in the external force acting on the hand. This figure has also been drawn so that the length of the moment arm is proportional to the actual length cited in the text (scale, 1:5).

Discussion Questions

1. If the patient in Figure 5.2A holds the lower handgrips while performing the push-up, will the torque counteracted by the elbow extensor muscles increase or decrease?

2. If the patient in Figure 5.2B increases the distance between his hands, how is the torque that is counteracted by the elbow extensor muscles affected?

3. Assume that the torque produced by the elbow extensor muscles is the same for the sitting and the prone push-up. It is known experimentally that a push-up is harder to execute in the prone position. Give an explanation for this fact.

5.3. Biceps and Triceps Muscle Strengthening Exercises against Resistance

Moment Equilibrium; Force Equilibrium

The purpose of this example is to determine whether the joint reaction force (with the torque held constant) is appreciably affected when the point of application of an external force is changed. Problems 5.3A and B involve strengthening exercises for the biceps muscle. The external force is applied at the wrist in problem 5.3A and at the middle of the forearm in problem 5.3B. The torque is kept constant, and the magnitude

of the joint reaction force is calculated in the two cases. Problems 5.3C
and D examine the same two situations during triceps muscle strengthening exercises.

Problems

5.3A. During isometric exercises for strengthening the biceps muscle, an
external force of 200 N is applied perpendicular to the forearm 0.3
m from the center of rotation of the elbow joint (Figure 5.3A). The
elbow is flexed 90° and supinated. The force produced by the biceps
muscle, which is assumed to be the only active muscle, has a
moment arm of 0.04 m and a line of application parallel to the long
axis of the humerus.

How large a force (B) must the biceps muscle generate to counteract
this torque? Disregard the weight of the forearm.

How large is the joint reaction force (R)?

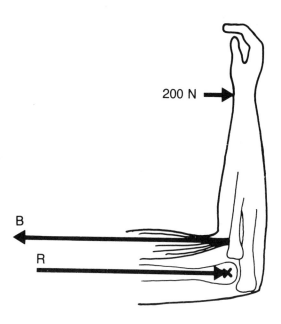

Figure 5.3A. Forces acting on the forearm during an isometric exercise for strengthening
the biceps muscle. An external force of 200 N is applied at the wrist. The force produced
by the biceps muscle (B) and the external force (200 N) give rise to the joint reaction force
(R) in the humeroulnar joint. The weight of the forearm can be disregarded.

5.3B. The external force (F) is now applied at the middle of the forearm,
0.15 m from the center of rotation of the elbow joint (Figure 5.3B).

How large must force F be for the magnitude of the biceps muscle
force (B) to equal that in problem 5.3A?

How large is the joint reaction force (R)?

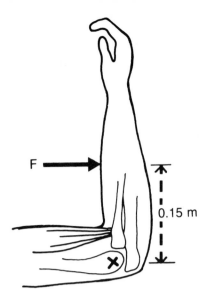

Figure 5.3B. During an isometric exercise for strengthening the biceps muscle, an external force (F) is applied 0.15 m from the center of rotation of the elbow joint. The weight of the forearm can be disregarded.

5.3C. During isometric exercises for strengthening the triceps muscle, an external force of 100 N is applied perpendicular to the forearm 0.3 m from the center of rotation of the elbow joint (Figure 5.3C). The elbow is flexed 90°. The force produced by the triceps muscle is assumed to act parallel to the long axis of the humerus. Its moment arm is 0.02 m. The weight of the forearm is disregarded.

How large a force (T) must the triceps muscle generate to counteract this torque?

How large is the joint reaction force (R)?

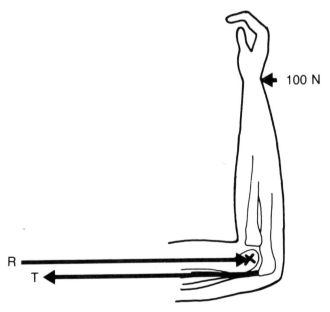

Figure 5.3C. The forces acting on the forearm during an isometric exercise for strengthening the triceps muscle. An external force of 100 N is applied at the wrist. The triceps muscle force (T) and the external force (100 N) give rise to the joint reaction force (R). The weight of the forearm can be disregarded.

5.3D. The external force (F) is now applied at the middle of the forearm, 0.15 m from the center of rotation of the elbow joint (Figure 5.3D).

How large must force F be for the magnitude of the triceps muscle force to equal that in problem 5.3C?
How large is the joint reaction force (R)?

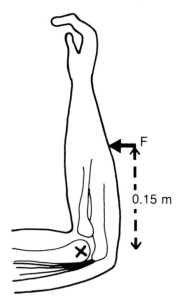

Figure 5.3D. During an isometric strengthening exercise for the triceps muscle, an external force (F) is applied to the forearm 0.15 m from the center of rotation of the elbow joint.

Discussion Questions

1. Why does the biceps muscle force equal the triceps muscle force during the two strengthening exercises even though the external torque is larger during the biceps muscle strengthening exercise?
2. Is the magnitude of the joint reaction force affected so greatly by the point of application of an external force that in practice it becomes important where this external force is applied?

5.4. Answers, Solutions, and Discussion

Answers to the Problems

5.1A.	120° 13.0±0.5 Nm	5.3A.	B=1500 N
	90° 15 Nm		R=1300 N
	30° 7.5±0.5 Nm		
	0° 0 Nm	5.3B.	F=400 N
			R=1100 N
5.1B.	90° 0 Nm		
	30° 13±1 Nm	5.3C.	T=1500 N
	0° 15 Nm		R=1600 N
5.2A.	M=15 Nm	5.3D.	F=200 N
			R=1700 N
5.2B.	M=16 Nm		

The ranges indicated in some of these answers allow for a reasonable margin of error in the graphic solution of the problems. These ranges were established on the basis that 1 cm in the figures corresponds to 5 cm in reality and that the maximum allowable error in measurement is 1 mm of length and 1° of angle.

Solution to Problem 5.1A

Find the torque (M) produced by the weight of the dumbbell at 120°, 90°, 30°, and 0° of elbow flexion. (The solution is given for 120° of elbow flexion only.)

Graphic Solution

The weight of the dumbbell (50 N) acts vertically. Its moment arm (a), the perpendicular distance from the weight of the dumbbell to the center of rotation of the elbow joint, is obtained by constructing a right triangle with a hypotenuse of 0.30 m and an acute angle of 30° (Figure 5.4.1).

$a = 0.26 \pm 0.01$ m

$M = 50 \cdot 0.26$

$M = 13 \pm 0.5$ Nm

Mathematical Solution

The length of the moment arm (a) of the weight of the dumbbell is obtained from Figure 5.4.1.

$a = 0.3 \cos 30°$

$a = 0.26$ m

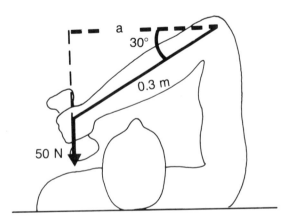

Figure 5.4.1. Dumbbell exercise performed by a supine patient with the elbow flexed 120°. The moment arm (a) of the weight of the dumbbell (50 N) relative to the center of rotation of the elbow joint forms one side of a right triangle with a hypotenuse of 0.30 m and an acute angle of 30°. The moment arm (a) is 0.26 m.

Answers to Discussion Questions in 5.1

1. In these two problems, the length of the external moment arm changes because the point of application of the force produced by the dumbbell shifts relative to the center of rotation of the elbow joint at different joint angles. Hence, the magnitude of the external torque is affected. The length of the moment arm, and thus the magnitude of the external torque, can also be affected by a change in the direction of the force. In these two problems, however, the direction does not change.

2. When exercises are performed with weights, a maximum torque cannot be produced at every point throughout the range of joint motion. The maximum external torque—and in general the maximum internal one as well—changes with the joint angle. This means that at best the torque will be maximum along only some part of the range of joint motion. Thus, exercise can be made more effective in different parts of the range of motion by altering the patient's position. For the patient in example 5.1, a 5-kg dumbbell does not produce a maximum torque at any angle of the elbow joint. When the patient is supine, the resistance is greatest at 90° of elbow flexion. When the patient is prone, the resistance is greatest at full extension of the elbow.

Solutions to Problems 5.2A and B

5.2A. Find the external torque (M) on the elbow joint during a sitting push-up.

The patient weighs 70 kg. His heels produce a force of 100 N against the force plate. The remaining 600 N is distributed equally between his

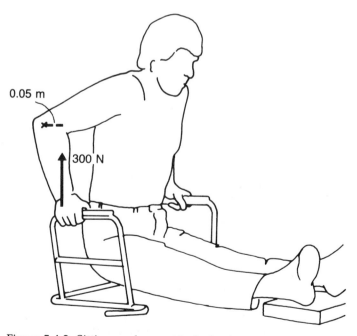

Figure 5.4.2. Sitting push-up with the heels supported. A vertical force of 300 N acts on the right hand. The moment arm of the force relative to the center of rotation of the elbow joint is marked with a broken line.

two hands. Figure 5.4.2 graphically illustrates the line of application, point of application, and direction of the reaction force on the right hand during a sitting push-up.

$$M = 0.05 \cdot \frac{700 - 100}{2}$$

$$M = 15 \text{ Nm}$$

5.2B. Find the external torque (M) on the elbow joint during a push-up in the prone position. Figure 5.4.3 graphically illustrates the reaction force on the right hand during this push-up.

$$M = 200 \cdot 0.08$$

$$M = 16 \text{ Nm}$$

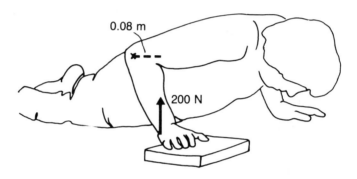

Figure 5.4.3. Push-up in the prone position with the lower legs supported. A vertical force of 200 N acts on the right hand. The moment arm of the force relative to the center of rotation of the elbow joint is marked with a broken line.

Answers to Discussion Questions in 5.2

1. If the patient in Figure 5.2A performs the push-up with his hands on the lower handgrips, the torque on the elbow joint will decrease because the moment arm of the vertical force acting on the hand will be shorter.

Note that in the sitting push-up the body may be raised up completely without activating the elbow extensor muscles if the line of application of the force on the hand shifts posterior to the transverse axis of rotation of the elbow joint. In this case, the elbow is "locked" in that the shoulder is rotated outward and the elbow is extended. The push-up is executed by depressing the shoulders (Perry, 1978).

2. During the push-up in a prone position, the torque that the elbow extensor muscles must counteract can be reduced by moving the hands farther apart. If the line of application of the force on the hand passes through the axis of rotation of the elbow joint, no torque is exerted on the joint. Thus, patients with weakness in the elbow extensor muscles

can perform this exercise more easily by keeping the hands far apart. At the same time, however, the torque on the shoulder joint increases.

3. Motion in the shoulder joint during a push-up in the prone position is not the same as the motion during a sitting push-up. Horizontal adduction takes place during the push-up in the prone position, whereas shoulder depression predominates during the sitting push-up. Shoulder adduction involves fewer muscles than does shoulder depression, which may explain why a push-up is more difficult from a prone position than from a sitting position.

Solutions to Problems 5.3A and C

5.3A. Find the muscle force (B) produced by the biceps muscle during strengthening exercises with an external force of 200 N applied at the wrist. In this problem, the clockwise moment is positive.

$\Sigma M = 0$ gives the magnitude of the biceps muscle force (B).

$200 \cdot 0.3 - B \cdot 0.04 = 0$

$B = 1500$ N

Find the joint reaction force (R) on the humeroulnar joint.
In this problem, the forces directed to the right are positive.

$\Sigma F = 0$ gives the magnitude of the joint reaction force (R).

$R + 200 - 1500 = 0$

$R = 1300$ N

5.3C. Find the muscle force (T) produced by the triceps muscle during strengthening exercises with an external force of 100 N applied at the wrist. In this problem, the counterclockwise moment is positive.

$\Sigma M = 0$ gives the magnitude of the muscle force (T).

$100 \cdot 0.3 - T \cdot 0.02 = 0$

$T = 1500$ N

Find the joint reaction force (R) on the humeroulnar joint.
In this problem, the force directed to the right is positive.

$\Sigma F = 0$ gives the magnitude of the joint reaction force.

$R - 100 - 1500 = 0$

$R = 1600$ N

Answers to Discussion Questions in 5.3

1. The external torque is 60 Nm during the biceps muscle strengthening exercise as opposed to 30 Nm during the triceps muscle exercise. Nevertheless, the force produced by the triceps muscle equals that produced by the biceps muscle because the triceps muscle force has a substantially shorter moment arm.

2. In practice, the magnitude of the joint reaction force is not appreciably affected by the point of application of the external force, i.e., whether the force is applied close to or at a distance from the joint. The decisive factor for the magnitude of the joint reaction force is the magnitude of the external torque. A large external torque is counteracted by a large muscle force, which in turn produces a large joint reaction force.

5.5. Commentary

The Humeroulnar Joint Reaction Force

In problem 5.3C, the humeroulnar joint reaction force is 1600 N when an external force of 100 N is applied at the wrist. The magnitude of the joint reaction force corresponds to about twice the body weight. Nicol et al. (1977) calculated that the load on the elbow on rising from a chair is 1700 N—about two times body weight—and when eating and dressing, about one-half body weight. The calculations were based on data from three young men weighing 65 to 75 kg. These data show that even the joints of the upper extremities are exposed to heavy loads during the activities of daily living.

Assumptions Made in Example 5.3

1. The biceps muscle force is assumed to counteract the extending torque on the elbow joint. Although relatively broad, this simplification does not affect too greatly the magnitude of the calculated muscle force and joint reaction force. The three most important elbow flexor muscles are the biceps muscle, the brachialis muscle, and the brachioradialis muscle. The brachialis muscle is activated at all times, regardless of the position of the forearm; the biceps muscle is most activated when the forearm is supinated, and the brachioradialis muscle is most activated with the forearm pronated (Basmajian, 1974). In example 5.3, the forearm is supinated. Hence, the brachioradialis muscle force can be disregarded. The direction of the brachialis muscle force is approximately the same as that of the biceps muscle force. These muscles can therefore be considered as a unit.

2. The moment arm of the force produced by the biceps muscle is assumed to be 4 cm. Bankov and Jorgensen (1969) found that the average lengths of the moment arms of the brachialis muscle force and the biceps muscle force were 3.2 and 4.4 cm, respectively.

3. The moment arm of the triceps muscle force is assumed to be 2 cm. Williams and Lissner (1977) cited a moment arm length of 2.5 cm for the triceps muscle force.

6

The Shoulder

The shoulder is composed of three synovial joints: the glenohumeral, the acromioclavicular, and the sternoclavicular. The scapulothoracic and the subacromial junctions also function as joints in the shoulder complex. The glenohumeral joint is a ball-and-socket joint with the center of rotation in the humeral head (Poppen and Walker, 1976). In example 6.1, motion of the glenohumeral joint is analyzed in the sagittal plane, and in example 6.2, in the frontal plane.

6.1. Strengthening the Shoulder Flexor Muscles with Pulleys

Torque

The following problems, based on two articles on pulley systems (Ekholm et al., 1978; Arborelius and Ekholm, 1978), show how the magnitude of the external torque changes as the glenohumeral joint moves through its range of motion during exercises with pulleys. These exercises are dynamic exercises, but if performed slowly and at a uniform speed, the laws of statics can be applied.

Problems

A patient is exercising the shoulder flexor muscles with the pulleys (Figure 6.1). The lower pulley is placed slightly below the surface of the table on which the patient is lying. The weight attached to the pulley rope exerts a force of 40 N. The distance from the center of rotation of the shoulder to the point of grip in the hand is 0.65 m. The weight of the arm is 35 N, and its center of gravity is 0.30 m from the center of rotation of the shoulder. The pulley can be considered to be friction-free.

6.1A. How large a torque (M) must the shoulder flexor muscles overcome at the beginning of the range of motion, at 10° of shoulder flexion? The pulley rope forms an angle of 160° with the long axis of the humerus (Figure 6.1A).

6.1B. How large a torque (M) is produced on the shoulder joint just before the middle of the range of motion, at 60° of shoulder flexion? The pulley rope forms an angle of 90° with the long axis of the humerus (Figure 6.1B.)

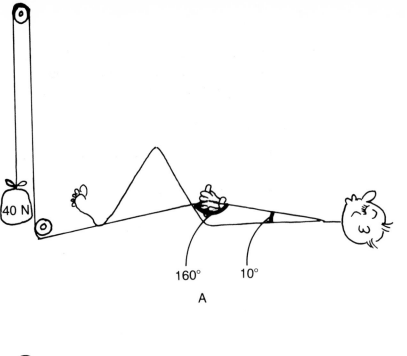

160° 10°

A

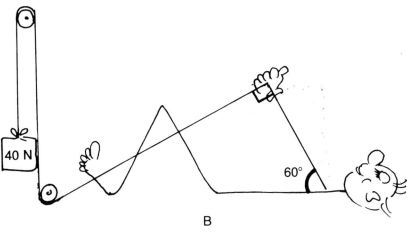

60°

B

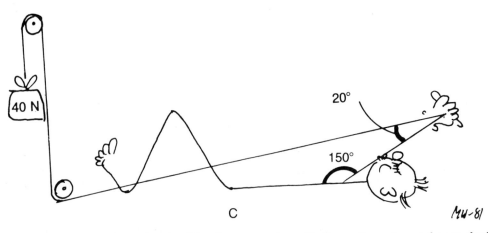

20°

150°

C

MU-81

Figure 6.1. Exercising the shoulder flexor muscles with the pulleys. A weight attached to the pulley rope exerts a force of 40 N. A. Shoulder flexed 10°. The pulley rope forms a 160° angle with the long axis of the humerus. B. Shoulder flexed 60°. The pulley rope forms a 90° angle with the long axis of the humerus. C. Shoulder flexed 150°. The pulley rope forms a 20° angle with the long axis of the humerus.

6.1C. How large a torque (M) is produced on the shoulder joint at 150° of shoulder flexion? The pulley rope forms an angle of 20° with the long axis of the humerus (Figure 6.1C).

Discussion Questions

1. At what angle of the shoulder joint is the maximum torque produced? Why?

2. Does the maximum torque occur at the same joint angle if the patient is moved closer to the pulley system? If the lower pulley is placed farther below the surface of the table on which the patient is lying?

6.2. The Joint Reaction Force during Shoulder Abduction with the Arm Extended and Flexed

Moment Equilibrium; Force Equilibrium

This example involves a comparison of the joint reaction force on an abducted shoulder joint with the arm extended and flexed. The effect of the length of the external moment arm on the magnitude of the torque, and hence on the joint reaction force, is illustrated. The data on the direction of the deltoid muscle force and the distance from the point of muscle attachment to the center of rotation of the shoulder joint were obtained from Williams and Lissner (1977).

Problems

6.2A. A patient with rheumatoid arthritis and shoulder pain is exercising her shoulder to prevent contractures and is instructed to abduct her arm with the elbow flexed instead of extended (Figure 6.2A). To what extent is the torque on the glenohumeral joint (M) reduced when the elbow is flexed rather than extended? The patient's body segment weights and their moment arms relative to the center of rotation of the shoulder joint are shown in the following table:

Body Segment	Weight (N)	Moment Arm (m)	
		Extended	Flexed
Upper arm	18	0.13	0.13
Forearm	10	0.36	0.18
Hand	4	0.59	0.00

6.2B. To what extent is the glenohumeral joint reaction force (R) reduced in this patient when the elbow is flexed rather than extended?

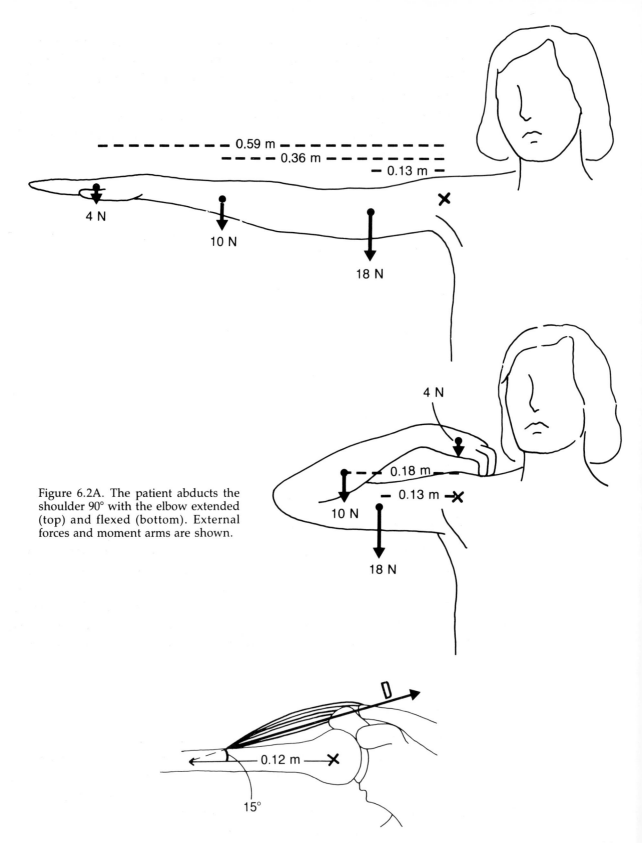

Figure 6.2A. The patient abducts the shoulder 90° with the elbow extended (top) and flexed (bottom). External forces and moment arms are shown.

Figure 6.2B. The position of the deltoid muscle relative to the humerus at 90° of shoulder abduction. The deltoid muscle force (D) forms a 15° angle with the long axis of the humerus, and the distance from the intersection point to the center of rotation of the shoulder joint is 0.12 m.

The deltoid muscle is assumed to be the only active muscle and to absorb the entire adducting torque produced by the weight of the arm. The deltoid muscle force (D) forms an angle of 15° with the long axis of the humerus. The distance from the center of rotation of the shoulder joint to the point of intersection of force D and the long axis of the humerus is 0.12 m (Figure 6.2B). Use necessary data from Figure 6.2A to solve the problem.

Discussion Question

Why is it difficult to make accurate biomechanical calculations when the shoulder joint is involved?

6.3. Answers, Solutions, and Discussion

Answers to the Problems

6.1A. 19 ± 2 Nm

6.1B. 31 ± 2 Nm

6.1C. 0 ± 2 Nm

6.2A. External torque (M) reduced by 4.2 Nm

6.2B. Glenohumeral joint reaction force (R) reduced by 131 ± 60 N

The ranges indicated in these answers allow for a reasonable margin of error in the graphic solution of the problems. In problems 6.1A–C, the ranges were established on the basis that 1 cm in the figure corresponds to 10 cm in reality. In problems 6.2A and B, 2 cm in the figure corresponds to 10 cm in reality. In all of the problems, 5 cm in the figures corresponds to 100 N. The maximum allowable error in measurement is 1 mm of length and 1° of angle.

Solution to Problem 6.1A

Find the external torque (M) on the glenohumeral joint.

Let a represent the moment arm of the force of the weight attached to the pulley (40 N), transmitted through the rope, and let b represent the moment arm of the weight of the arm (35 N). The patient sustains the entire force of 40 N, since the pulleys are assumed to be free of friction. In this problem, the counterclockwise moments are positive.

$$M = (40a + 35b)Nm$$

Graphic Solution

The moment arm a is obtained by constructing a right triangle with a hypotenuse of 0.65 m (the length of the arm from the center of rotation of the shoulder joint to the point of grip in the hand) and one acute angle of 20° (the supplementary angle of the 160° angle given in the problem) (Figure 6.3.1). Side a is opposite the 20° angle. The length of a is measured.

$a = 0.22 \pm 0.02$ m

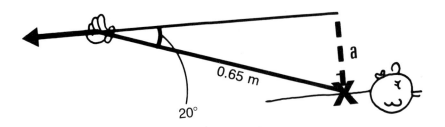

Figure 6.3.1. Exercising the shoulder flexor muscles with the pulleys. The moment arm (a) of the force transmitted through the pulley rope forms the side of a right triangle with a hypotenuse of 0.65 m and one acute angle of 20°.

Moment arm b is obtained in the same way. A right triangle with a hypotenuse of 0.3 m and one acute angle of 10° is constructed (Figure 6.3.2).

Side b is the side adjacent to the 10° angle. The length of b is measured.

$b = 0.30 \pm 0.02$ m

The values for moment arms a and b are substituted into the above equation.

$M = 40 \cdot 0.22 + 35 \cdot 0.30$

$M = 19 \pm 2$ Nm

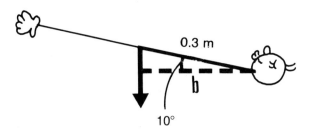

Figure 6.3.2. The patient is supine with the shoulder flexed 10°. The moment arm (b) of the weight of the arm relative to the center of rotation of the shoulder joint forms the side of a right triangle with a hypotenuse of 0.3 m and one acute angle of 10°.

<table>
<tr><td>**Mathematical
Solution**</td><td>The value for moment arm a of the force transmitted through the pulley rope is obtained from Figure 6.3.1.</td></tr>
</table>

$a = 0.65 \sin 20°$

The value for moment arm b of the weight of the arm is obtained from Figure 6.3.2.

$b = 0.30 \cos 10°$

$M = 40 \cdot 0.65 \sin 20° + 35 \cdot 0.30 \cos 10°$

$M = 19$ Nm

Answers to Discussion Questions in 6.1

1. In this example the maximum torque is produced when the shoulder is flexed 60°. The force transmitted through the pulley rope is perpendicular to the long axis of the humerus at that joint angle, and the external moment arm is therefore the longest.

2. Two factors determine where in the range of shoulder motion the torque produced by the pulley system is maximal: the distance of the shoulder joint from the pulley system, and the height of the lower pulley relative to the shoulder (because that pulley determines the direction of the pulley rope). If the patient is moved closer to the pulley system, the maximum external torque occurs earlier in the range of motion. The same is true if the lower pulley is lowered farther.

Solutions to Problems 6.2A and B

6.2A. Find the external torque (M) on the glenohumeral joint with the shoulder abducted 90° and the elbow flexed and extended. (The solution is given for elbow extension only.)

In this problem, the counterclockwise moments are positive.

$M = 18 \cdot 0.13 + 10 \cdot 0.36 + 4 \cdot 0.59$

$M = 8.3$ Nm

The external torque (M) is reduced by the following amount when the elbow is flexed rather than extended:

$8.3 - 4.1 = 4.2$ Nm

6.2B. Find the glenohumeral joint reaction force (R) with the shoulder abducted 90° and the elbow extended.

Graphic Solution   In this problem, the forces acting downward are positive. The following forces act on the arm: its own weight (18 N + 10 N + 4 N) acting vertically downward, the deltoid muscle force (D) acting at a 15° angle to the horizontal plane, and the joint reaction force (R).

Let the moment arm of force D be d.

$\Sigma M = 0$ gives the magnitude of the deltoid muscle force D.

$Dd = 8.3$ Nm

Moment arm d is obtained graphically by constructing a right triangle with a hypotenuse of 0.12 m and an acute angle opposite d of 15° (Figure 6.3.3). The length of d is measured.

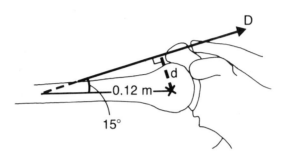

Figure 6.3.3. Right triangle with a 15° acute angle and a hypotenuse of 0.12 m. Side d represents the moment arm of the deltoid muscle force (D).

$d = 0.031 \pm 0.003$ m

$D \cdot 0.031 = 8.3$ Nm

$D = 267 \pm 30$ N

$\Sigma F = 0$ gives the joint reaction force R.

The magnitude and direction of R are obtained by means of the polygon method (Figure 6.3.4).

$R = 260 \pm 35$ N

Find the joint reaction force (R) with the shoulder abducted 90° and the elbow flexed. The solution is similar to that given for elbow extension.

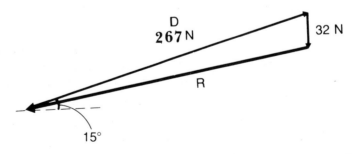

Figure 6.3.4. The polygon of forces reveals the glenohumeral joint reaction force (R) during shoulder abduction with the arm extended. The deltoid muscle force (D) (267 N) forms a 15° angle with the horizontal plane. The weight of the arm (32 N) acts vertically downward.

M = 4.1 Nm

D = 133 ± 15 N

R = 129 ± 25 N

With the shoulder abducted 90°, the joint reaction force (R) is reduced by the following amount when the elbow is flexed rather than extended:

260 ± 35 N − 129 ± 25 N = 131 ± 60 N

The maximum allowable error in the graphic solution for force R is 35 N + 25 N = 60 N (the sum of the upper limits for the force components).

Mathematical Solution

The moment arm (d) of the deltoid muscle force (D) and the magnitude of this force with the shoulder abducted 90° and the elbow extended are calculated with the use of data from Figure 6.3.3.

d = 0.12 sin 15°

d = 0.031 m

D · 0.031 − 8.3 = 0

D = 267 N

The joint reaction force (R) is calculated with the use of data from Figure 6.3.4 by means of the cosine theorem.

$R^2 = 267^2 + 32^2 − 2 \cdot 267 \cdot 32 \cdot \cos (90° − 15°)$

R = 262 N

Answer to Discussion Question in 6.2

Making accurate biomechanical calculations for the shoulder joint is complicated because the anatomy of this joint is complex. Shoulder joint motion takes place in all three planes, and there is considerable activity in many muscles other than the deltoid muscle, for example, the supraspinatus muscle and the other rotator cuff muscles. The assumptions made in the simplified calculations in these problems are too broad to yield accurate values. The solutions to the problems yield absolute minimum values for the joint reaction force, since no consideration is given to the forces produced by the muscles and ligaments that stabilize the joint.

The purpose of the example is simply to show that a reduced external torque results in a smaller joint reaction force. In this case, the joint reaction force decreases by 50% (from about 260 N to about 130 N) when the elbow is flexed rather than extended. Poppen and Walker (1978) calculated the glenohumeral joint reaction force to be about 0.9 times body weight with the shoulder abducted 90° in the scapular plane. Activity in the deltoid, supraspinatus, infraspinatus, and latissimus dorsi muscles was recorded and included in the calculations of the joint reaction force.

7

The Hip

The hip is a ball-and-socket joint with the center of rotation in the femoral head. Examples 7.1 and 7.2 consider the hip joint in the sagittal plane. Examples 7.3 through 7.7 analyze the influence of the abductor muscle force on loading of the hip joint in the frontal plane. Four examples involving the abductor muscle group have been included because, from a clinical standpoint, this muscle group is an important stabilizer of the hip joint (Blount, 1956). Also, during walking the torque on the hip joint is largest in the frontal plane (Johnston et al., 1979). The loads on the hip joint during slow walking are comparable to those produced during a one-leg stance (McLeish and Charnley, 1970). Example 7.8 analyzes the bending moment that can arise in a total hip joint prosthesis.

7.1. Stretching the Hamstring Muscles Passively in Two Different Ways

Torque

In this example, the flexing torque on the hip joint produced during two different methods of passively stretching shortened hamstring muscles is compared.

Problems

A paraplegic patient is stretching the hamstring muscles by bending forward from a sitting position (Figure 7.1A). A physical therapist comes to assist him. She places him on his back, flexes the hip joint with the knee extended, and applies a force of 100 N to the ankle perpendicular to the long axis of the leg (Figure 7.1B).

7.1A. The patient is sitting with the upper body flexed as far as possible (Figure 7.1A). The weight of the head and trunk is 400 N. The center of gravity of the head and trunk is 0.2 m anterior to the center of rotation of the hip. The combined weight of the arms is 80 N, and their common center of gravity is 0.5 m anterior to the center of rotation of the hip. How large is the flexing torque (M) on the hip?

7.1B. The therapist applies a force of 100 N to the patient's ankle, perpendicular to the long axis of the leg (Figure 7.1B). The distance from the center of rotation of the hip to the point of application of

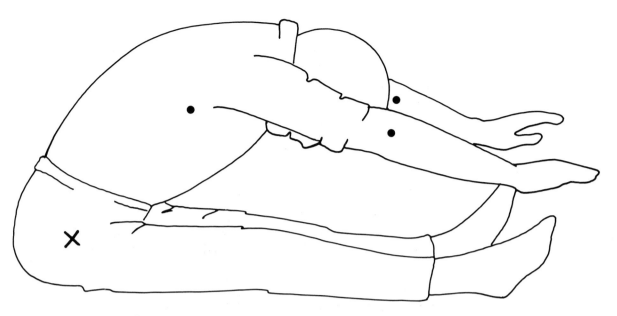

Figure 7.1A. The patient is sitting with the upper body flexed as far as possible. The center of rotation of the hip is marked with an X. The centers of gravity of the head and trunk, and of the arms, are marked with dots.

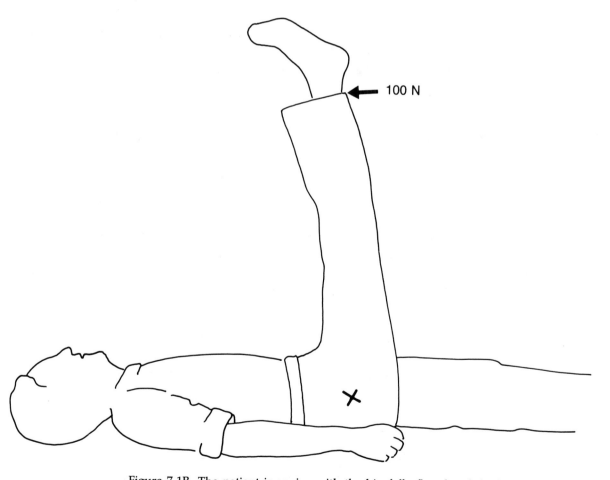

100 N

Figure 7.1B. The patient is supine with the hip fully flexed and the knee extended. A force of 100 N is applied to the ankle perpendicular to the long axis of the leg. The center of rotation of the hip is marked with an X.

the external force is 0.75 m. The weight of the leg can be disregarded, since its line of application passes approximately through the center of rotation of the hip. How large is the flexing torque (M) on the hip?

Discussion Questions

1. With the patient bending forward from the sitting position, would the torque on the hip joint increase or decrease if the hamstring muscles were shorter? The hamstring muscles are assumed to counteract the flexing torque.

2. When the hip is moved through its range of motion, how does the weight of the leg affect the flexing torque on the hip joint produced by a force applied at the ankle of the patient in the supine position with the knee extended? First consider the torque for the range of hip motion from 0° to 90°, and then from 90° to 120°. How is the magnitude of the external force that the physical therapist applies at the ankle affected throughout the range of hip motion?

7.2. Active Straight Leg Lifts Performed in the Supine Position

Torque

This example illustrates how several individual centers of gravity can be replaced by a single one. The example can be used in clinical discussions of joint loads produced by muscle activity in a supine patient.

Problems

A supine patient performs an active leg lift with the knee fully extended (Figure 7.2).

7.2A. How large is the torque (M) on the hip produced by the weight of the leg?

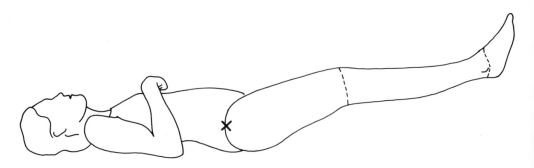

Figure 7.2. Active straight leg lift with the patient in the supine position. The center of rotation of the hip is marked with an X. The figure has been drawn so that the lengths of the moment arms are proportional to the actual lengths given in the text. (Scale, 1:12.)

Draw in the centers of gravity and the moment arms of the weights of the various body segments in Figure 7.2, using the following data for an individual weighing 70 kg:

Body Segment	Weight (N)	Moment Arm (m)
Thigh	67	0.18
Lower leg	33	0.60
Foot	10	0.90

7.2B. What is the horizontal distance (a) from the center of rotation of the hip to the common center of gravity of the three body segments? Use data from the table in problem 7.2A.

Discussion Questions

1. What data are lacking in the problems for the position of the common center of gravity of the thigh, lower leg, and foot in the sagittal plane to be designated precisely?

2. Develop the problems further and discuss the hip joint reaction force. How do the values for the hip joint reaction force during the active leg lift performed in a supine position compare with the values for the symmetrical standing position and for the one-leg stance (Trendelenburg test)?

7.3. Analysis of a One-Leg Stance Using Two Free Body Diagrams

Torque

This example is useful for reviewing free body analysis. It provides practice in determining the direction of given forces acting on a free body and the torque that they produce at a given point.

There is often discussion about whether the weight of the standing leg should be included in calculations of the torque and forces on the hip joint during a one-leg stance. The example illustrates that the result is the same whether the calculation takes into account the weight of the whole body or excludes the weight of the standing leg. Thus, either method of calculating the torque and forces on the hip joint is valid.

Problem

A patient is standing on one leg (Figure 7.3). The abductor muscles are assumed to absorb the entire adducting torque produced by the body weight on the hip joint of the standing leg. How large an abductor muscle force (A) is required to balance the upper body?

Work out two separate solutions, one in which the upper body constitutes the free body, and one in which the standing leg is the free body.

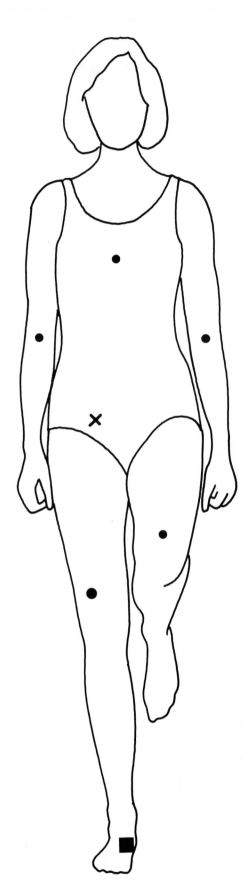

Figure 7.3. The patient is standing on one leg. The centers of gravity of the head and trunk, the arms, and the legs are marked with dots. The point of application of the ground reaction force against the foot is marked with a square. The center of rotation of the hip joint of the standing leg is marked with an X.

In Figure 7.3, draw in the forces acting on the two free bodies. Use the following data for an individual weighing 69 kg:

Body Segment	Weight (N)	Moment Arm (m)
Head and trunk	400	0.05
Right arm	35	0.09
Left arm	35	0.20
Right leg (standing leg)	110	0.01
Left leg (lifted leg)	110	0.15
Force		
Ground reaction force		0.06
Abductor muscle force		0.05

Discussion Question

Locate the common center of gravity of all of the individual body segments when the upper body is used as the free body; when the standing leg is used as the free body. Does the moment arm of the weight of the upper body differ from the moment arm of the ground reaction force?

7.4.–7.6. The Effect of Various Exercises and Treatment Methods on Loads on the Hip Joint

Moment Equilibrium; Force Equilibrium

The following three examples, which allow a comparison of the loads on the hip joint during two types of exercises and two methods of treatment, can be used as the basis for a discussion of various therapeutic modalities.

In example 7.4, the torque counteracted by the abductor muscles during a one-leg stance (the Trendelenburg test) is calculated, and the joint reaction force is then determined.

In example 7.5, the value for the torque on the hip joint during the one-leg stance is used to determine the magnitude of the external force that must be applied to the leg (at a certain distance from the center of rotation of the hip joint) during abductor muscle exercises performed in a supine position, and in a side-lying position, for the abductor muscle force to equal that produced during the one-leg stance. The effect of the exercise position on the joint reaction force is then analyzed.

In example 7.6, two methods for reducing the loads on the hip joint—cane use and weight loss—are examined.

The following data are required for solving the problems in examples 7.4 through 7.6:

Weight of the patient	820 N
Weight of the leg	120 N
Length of the leg (from the center of rotation of the hip to the ankle joint)	0.80 m
Moment arm of the abductor muscle force	0.05 m
Moment arm of the weight of the leg with the patient in the supine position	0.35 m
Moment arm of the weight of the body, excluding the standing leg	0.10 m

The line of application of the abductor muscle force (A) forms a 70° angle with a line passing through both trochanters in the transverse plane (see Figure 7.4B).

7.4. Hip Joint Loading during the One-Leg Stance (Trendelenburg Test)

Problems

7.4A. How large a torque (M) must the abductor muscles counteract when a patient stands on one leg (the Trendelenburg test) and balances the upper body in the frontal plane (Figure 7.4A)? The abductor muscles are assumed to absorb the entire torque.

7.4B. How large is the joint reaction force (R) on the hip joint, and what is its direction (α) relative to the transverse plane? Draw in the forces acting on the hip joint in Figure 7.4B.

7.5. Hip Joint Loading during Abductor Muscle Exercises Performed in the Supine Position

Problems

7.5A. The same patient presented in example 7.4 is exercising the abductor muscles isometrically in the supine position with the hip joint in full extension. The friction of the leg against the exercise table can be disregarded.

How large an external force (F) must be applied at the ankle (perpendicular to the long axis of the tibia) for the abductor muscle force to equal that in example 7.4?

How large will the hip joint reaction force (R) be and in what direction (α) will it act relative to the transverse plane?

7.5B. Instead of lying supine, the patient is lying on his side with the hip joint fully extended.

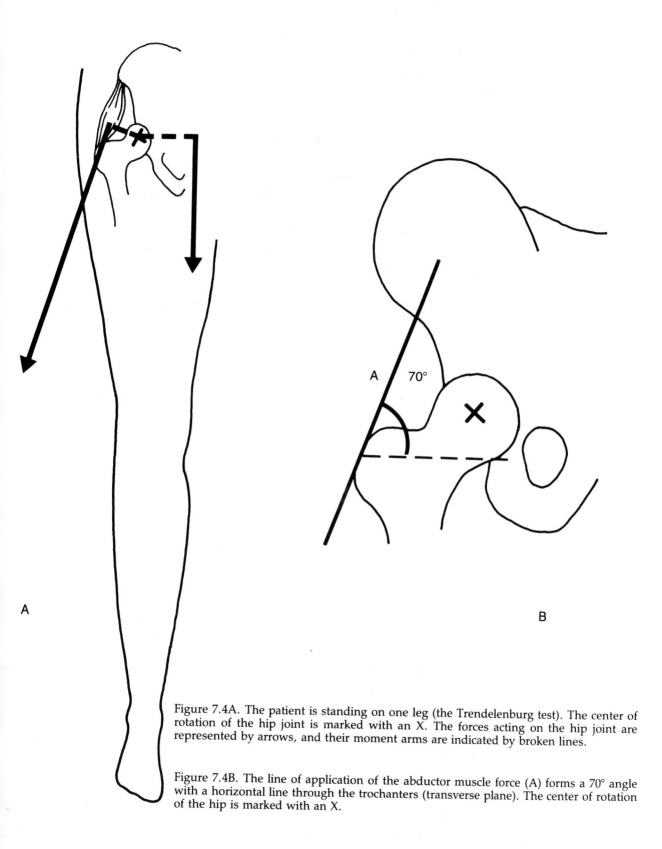

Figure 7.4A. The patient is standing on one leg (the Trendelenburg test). The center of rotation of the hip joint is marked with an X. The forces acting on the hip joint are represented by arrows, and their moment arms are indicated by broken lines.

Figure 7.4B. The line of application of the abductor muscle force (A) forms a 70° angle with a horizontal line through the trochanters (transverse plane). The center of rotation of the hip is marked with an X.

How large an external force (F) must be applied at the ankle (perpendicular to the long axis of the tibia) for the abductor muscle force to equal that in example 7.4?

How large is the joint reaction force (R) on the hip joint and in what direction (α) does it act relative to the horizontal plane?

7.6. The Effect of Cane Use and Weight Loss on Hip Joint Loading

Problems

7.6A. To relieve pain in the hip joint, the patient, who is standing on one leg, uses a cane in the hand on the side opposite the standing leg. The moment arm of the force through the cane is 0.30 m.

How large a torque (M) must the abductor muscles counteract to balance the upper body if the patient exerts a vertical load of 150 N on the cane?

How large is the joint reaction force (R) on the hip joint and in what direction (α) does it act relative to the transverse plane? Use Figure 7.4A and data given in the introduction to examples 7.4 through 7.6 to solve the problem.

7.6B. By what amount are the torque (M) and the abductor muscle force (A) reduced if the same patient loses 15 kg and no longer uses a cane? The patient is "top-heavy," and the leg weight is assumed to be the same after the weight loss.

How large is the joint reaction force (R) on the hip joint and in what direction (α) does it act relative to the transverse plane? Use data given in the introduction to examples 7.4 through 7.6 to solve the problem.

Discussion Questions for 7.4 through 7.6

1. How large is the adducting torque on the hip joint produced by the weight of the upper body during the one-leg stance (the Trendelenburg test) for a person weighing 82 kg? What factors determine its magnitude?

2. How does the ratio between the moment arms of the abductor muscle force and the body weight affect the magnitude of the joint reaction force during the Trendelenburg test?

3. For the torque to correspond to that during the one-leg stance (the Trendelenburg test), how large an external force must be applied at the ankle when the patient lies on his back with the hip joint fully extended? When he lies on his side? Does the magnitude of the external force in both cases match your experience with abductor muscle exercises using weights and pulleys?

4. If the external torque is kept constant, is the joint reaction force affected when the patient shifts from a one-leg stance to a supine position?

5. Which has a greater effect on the hip joint reaction force, the loss of weight (15 kg) or the use of the cane?

7.7. Hip Joint Loading during Symmetrical and Asymmetrical Lifting

Moment Equilibrium; Force Equilibrium

This example illustrates the effect of the distribution of a burden on hip joint loading. The loading of the hip joint when a suitcase is carried in one hand is compared with the loading when the same type of suitcase is carried in each hand (in other words, when the burden is twice as heavy).

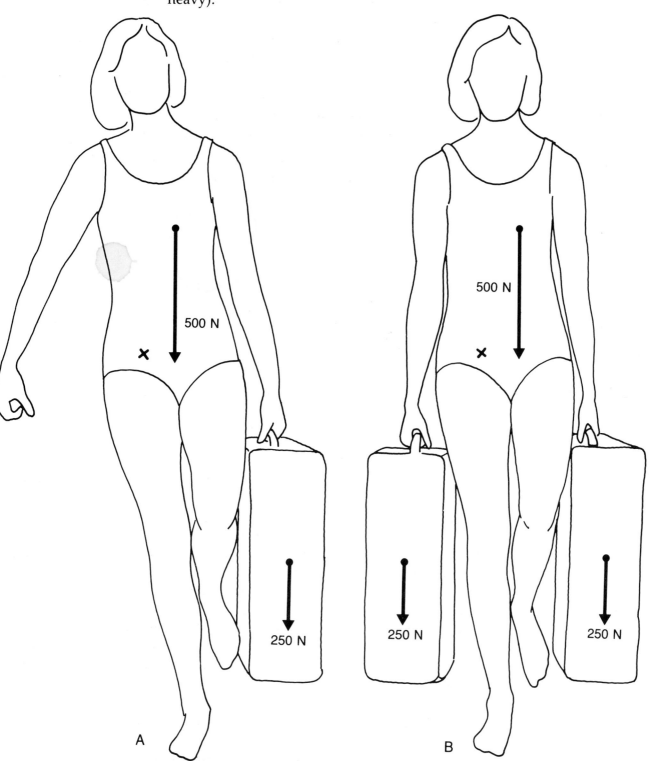

Figure 7.7. A person is standing on one leg while carrying a 25-kg suitcase in the hand on the side opposite the standing leg (Figure A) and a 25-kg suitcase in each hand (Figure B). The body weight excluding the weight of the standing leg is 500 N. The center of rotation of the hip joint is marked with an X.

Problems

7.7A. How large is the joint reaction force (R) in the right hip when a person stands on the right leg while carrying a 25-kg suitcase in the left hand (Figure 7.7A)? Use the following data:

Body weight	600 N
Weight of the standing leg	100 N
Moment arm of the weight of the suit-case	0.31 m
Moment arm of the abductor muscle force	0.05 m
Moment arm of of the body weight excluding the weight of the standing leg	0.06 m

The abductor muscle force (A) acts at a 70° angle to the transverse plane.

7.7B. How large is the joint reaction force (R) on the right hip joint when a person stands on the right leg while carrying a 25-kg suitcase in each hand (Figure 7.7B)?

The moment arm for the weight of the suitcase in the right hand is 0.14 m, and for the weight of the suitcase in the left hand, 0.31 m. The moment arms of the body weight and of the abductor muscle force are assumed to be the same as in problem 7.7A.

Discussion Question

Which produces a larger hip joint reaction force, carrying 25 kg in one hand or carrying 50 kg evenly distributed between two hands? Does the distribution of the burden have a sufficient effect on the joint reaction force to justify its mention during ergonomic counseling?

7.8. Loading of a Total Hip Joint Prosthesis Secured in a Varus and a Valgus Position

Torque

The example shows how the bending moment in the frontal plane, and hence the tensile stress, in the femoral portion of a total hip joint prosthesis differ when the device is secured to the bone in a varus position and when it is secured in a valgus position. The calculations of the bending moment relate to a point in the area wherein failure of the prosthesis is most likely to occur.

Problems

The following forces can be assumed to act on a total hip joint prosthesis: the joint reaction force (R) from the acetabulum, the supporting force (C) from the calcar region, the lateral force (L) from the femoral shaft, and a force (T) on the distal tip of the prosthesis (Figure 7.8).

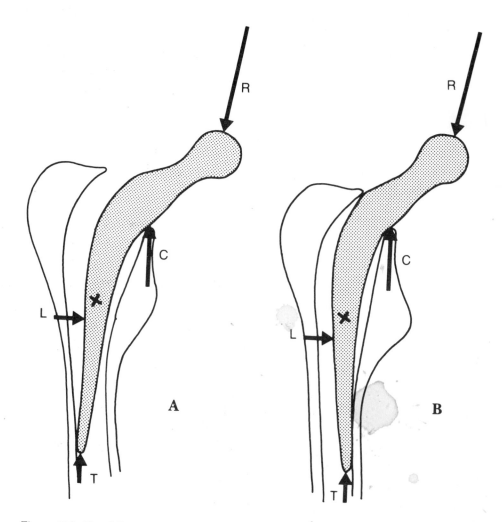

Figure 7.8. The following forces act on the total hip joint prosthesis: the joint reaction force (R) from the acetabulum, a lateral force (L) from the femur, a force (T) on the distal tip of the prosthesis, and a supporting force (C) from the calcar region. An arbitrarily selected center of rotation is marked with an X. The prosthesis is placed in the medullary canal in a varus position in Figure A and a valgus position in Figure B.

7.8A. During total hip joint replacement surgery, a prosthesis is secured to the bone in a varus position. How large is the bending moment (M) in the frontal plane at an arbitrarily selected center of rotation in the prosthesis (marked with an X in Figure 7.8A)? The joint reaction force (R) is 2800 N, and its moment arm relative to the center of rotation is 0.04 m. The supporting force C from the calcar region is 1400 N and has a moment arm of 0.02 m (Andriacchi, 1981).

7.8B. In another hip joint replacement procedure, a prosthesis is secured in a valgus position (Figure 7.8B). The moment arm of the joint reaction force R is 0.03 m. How large is the bending moment (M) in the frontal plane at the same center of rotation as in problem 7.8A? The magnitude of the supporting force C and the length of its moment arm remain unchanged.

Discussion Questions

1. How is the tensile stress in the lateral part of the prosthesis affected by varus and by valgus positioning of the prosthesis?

2. Cite some other factors that affect the bending moment in the frontal plane, and thus the occurrence of prosthesis failure. Try to give a mechanical explanation for each factor cited.

7.9. Answers, Solutions, and Discussion

Answers to the Problems

7.1A. 120 Nm

7.1B. 75 Nm

7.2A. 41 Nm

7.2B. 0.37 m

7.3. 806 or 807 N

7.4A. 70 Nm

7.4B. $R = 2067 \pm 25$ N
$\alpha = 77 \pm 1°$

7.5A. $F = 88$ N
$R = 1432 \pm 20$ N
$\alpha = 67 \pm 1°$

7.5B. $F = 35$ N
$R = 1460 \pm 20$ N
$\alpha = 64 \pm 1°$

7.6A. 25 Nm
$R = 1050 \pm 25$ N
$\alpha = 81 \pm 1°$

7.6B. Torque (M) reduced by 15 Nm
Abductor muscle force (A) reduced by 300 N
$R = 1630 \pm 25$ N
$\alpha = 77 \pm 1°$

7.7A. $R = 2870 \pm 30$ N

7.7B. $R = 2410 \pm 30$ N

7.8A. 84 Nm

7.8B. 56 Nm

The ranges indicated for some of these answers allow for a reasonable margin of error in the graphic solution of the problems. These ranges were established on the assumption that 1 cm in the figures corresponds to 100 N in reality (except in example 7.2, where 1 cm corresponds to 200 N) and that the maximum allowable error in measurement is 1 mm of length and 1° of angle.

7.1A. Find the flexing torque (M) on the hip joint when a patient passively stretches the hamstring muscles by flexing the upper body as far as possible in a sitting position (Figure 7.9.1).

M = 400 · 0.2 + 80 · 0.5

M = 120 Nm

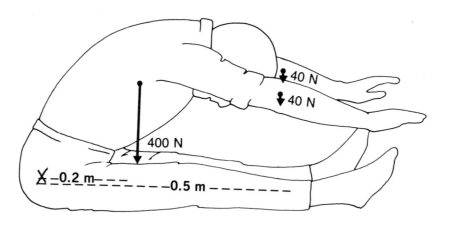

Figure 7.9.1. The patient bends forward as far as possible from a sitting position. The weight of the head and trunk is 400 N and has a moment arm of 0.2 m; the combined weight of the arms is 80 N, and the moment arm is 0.5 m.

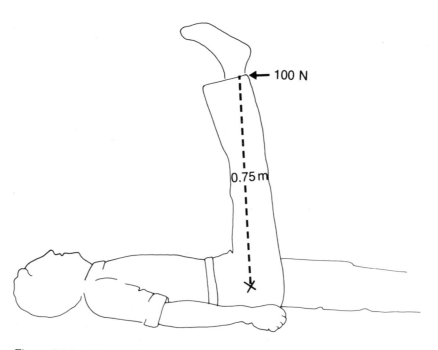

Figure 7.9.2. A force of 100 N with a moment arm of 0.75 m is applied proximally at the ankle of a supine patient with the hip fully flexed. The weight of the leg is disregarded because its line of application passes through the center of rotation of the hip.

7.1B. Find the flexing torque (M) on the hip joint when a therapist applies an external force of 100 N at the ankle of a supine patient whose hip is fully flexed and the knee extended (Figure 7.9.2).

$M = 100 \cdot 0.75$

$M = 75$ Nm

Answers to Discussion Questions in 7.1

1. If the hamstring muscles were shorter, the patient would not be able to lean as far forward, and the moment arms of the weight of the arms and the weight of the head and trunk would be shorter. Thus, the torque on the hip joint would decrease.

2. From 0 to 90° of hip flexion, the weight of the leg produces an extending torque on the hip joint; i.e., it counteracts the flexing torque produced by the external force applied by the therapist. The therapist must apply a greater external force because he or she has to overcome the extending torque of the weight of the leg before the passive structures can be stretched. At 90° of hip flexion, the weight of the leg produces no torque because its line of application passes through the center of rotation of the hip joint. With more than 90° of flexion, the weight of the leg produces a flexing torque, thus reducing the external force that the therapist must apply.

Solutions to Problems 7.2A and B

7.2A. Find the external torque (M) on the hip joint during an active leg lift with the patient supine and the knee fully extended.

$M = 67 \cdot 0.18 + 33 \cdot 0.60 + 10 \cdot 0.90$

$M = 41$ Nm

Figure 7.9.3 shows the weights of the individual body segments and the lengths of their moment arms relative to the center of rotation of the hip joint.

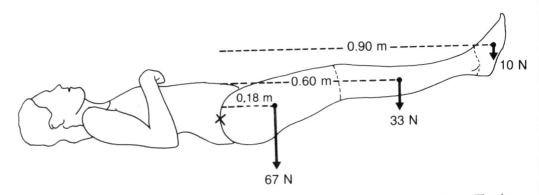

Figure 7.9.3. The patient performs an active leg lift in the supine position. The forces acting on the right hip joint, produced by the weight of the individual body segments, are represented by arrows, and their moment arms are indicated by broken lines. The center of rotation of the hip is marked with an X.

7.2B. Find the horizontal distance (a) from the center of rotation of the hip to the common center of gravity of the three body segments.

The horizontal distance (a) constitutes the moment arm for the entire weight of the leg (67 N + 33 N + 10 N).

$$a = \frac{41}{67 + 33 + 10}$$

$$a = 0.37 \text{ m}$$

Answers to Discussion Questions in 7.2

1. A precise designation of the position of the common center of gravity of the thigh, lower leg, and foot in the sagittal plane requires that it be located in both the craniocaudal and the ventrodorsal directions. In this example, only the craniocaudal direction is considered. Designation of the position of the center of gravity in the ventrodorsal direction would have been possible if the angle of flexion of the hip joint had been given.

2. During an active leg lift in a supine position, the hip flexor muscles counteract the torque produced by the weight of the leg (41 Nm in the example). If it is assumed that the moment arm of the hip flexor muscle force is 0.05 m, the muscle force is 820 N. This force determines the magnitude of the joint reaction force, which is about 25% larger than the body weight. The magnitude of this joint reaction force corresponds with intravital measurements and with experimentally verified biomechanical calculations (Rydell, 1966; Frankel, 1973). The hip joint reaction force during symmetrical standing is approximately one-third of body weight; during a one-leg stance, the joint reaction force is two to three times body weight (Rydell, 1966; Frankel and Nordin, 1980). The joint reaction force produced during a leg lift in the supine position is therefore larger than that produced during standing with the weight evenly distributed on both feet, but less than that produced during a one-leg stance.

Solution to Problem 7.3

Find the abductor muscle force (A) acting on the hip joint during a one-leg stance.

If the upper body is considered to be a free body (Figure 7.9.4, left), the following forces are present: the weight of the head and trunk, the weight of the arms, the weight of the lifted leg, and the abductor muscle force. In this problem, the clockwise moments are positive.

$\Sigma M = 0$ gives the magnitude of the abductor muscle force A.

$$400 \cdot 0.05 - 35 \cdot 0.09 + 35 \cdot 0.20 + 110 \cdot 0.15 - A \cdot 0.05 = 0$$

$$A = 807 \text{ N}$$

If the standing leg is considered to be a free body (Figure 7.9.4, right), the following forces act: the ground reaction force, the weight of the

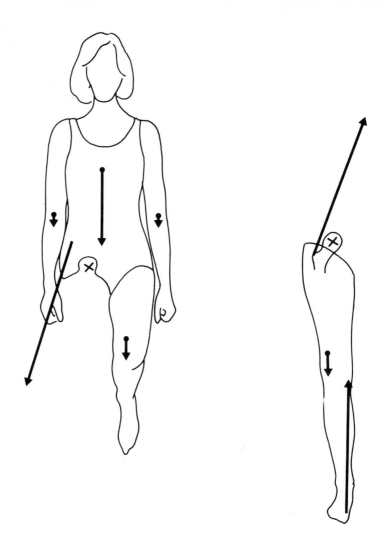

Figure 7.9.4. Free body diagrams for calculating the torque on the hip joint during a one-leg stance. Either the upper body (left figure) or the standing leg (right figure) can constitute a free body. The involved forces are represented by arrows.

standing leg, and the abductor muscle force. In this problem, the counterclockwise moment is positive.

$\Sigma M = 0$ gives the magnitude of the abductor muscle force A.

$690 \cdot 0.06 - 110 \cdot 0.01 - A \cdot 0.05 = 0$

$A = 806$ N

Answer to Discussion Question in 7.3

When the upper body is used as the free body, the weight of the standing leg is subtracted from the body weight. When the standing leg is used as the free body, the entire weight of the body is included. The common center of gravity of the arms, the head and trunk, and the lifted leg lies somewhat lateral to the center of gravity of the entire body. In example 7.3 this distance is 1 cm. Thus, the moment arm of the weight of the upper body is slightly smaller than the moment arm of the weight of the total body (equal to the ground reaction force).

In practice, the simplest way to calculate the torque on the hip joint is to consider the standing leg as the free body. The point of application, direction, and magnitude of the ground reaction force can be determined with the use of a force plate. Inman (1947) pointed out that during a one-leg stance, the center of gravity of the standing leg falls directly beneath the hip joint in the frontal plane, or at the most, 1.5 cm laterally or medially. Therefore, the torque on the hip joint produced by the standing leg in the frontal plane can be disregarded. In the example, this torque was included for the sake of completeness.

Discussions of the torque on the hip joint usually begin with a consideration of the torque produced by the portion of the body weight above the hip joint. This approach, which is illustrative and easy to comprehend, is defensible from a mechanical standpoint.

Solutions to Problems 7.4A and B, 7.5A, and 7.6A

7.4A. Find the external torque (M) on the hip joint during a one-leg stance with the upper body balanced in the frontal plane. The upper body constitutes the free body.

$$M = (820 - 120) \cdot 0.1$$

$$M = 70 \text{ Nm}$$

7.4B. Find the hip joint reaction force (R) in direction α relative to the transverse plane during a one-leg stance. The upper body constitutes the free body, and the following forces act on it: the body weight, excluding the weight of the standing leg (820 N − 120 N), acts vertically; the abductor muscle force (A) acts at a 70° angle to the transverse plane; and the line of application of the joint reaction force (R) passes through the center of rotation of the hip joint.

Graphic Solution

The abductor muscle force (A) is calculated first, and the joint reaction force (R) is then determined.

In this problem, the counterclockwise moment is positive.

$\Sigma M = 0$ gives the magnitude of the abductor muscle force A.

$$A \cdot 0.05 - 70 = 0$$

$$A = 1400 \text{ N}$$

$\Sigma F = 0$ gives the joint reaction force R.

Figure 7.9.5 shows the forces acting on the hip joint. The magnitude and direction of force R are obtained by means of the polygon method (Figure 7.9.6).

$$R = 2067 \pm 25 \text{ N}$$

$$\alpha = 77 \pm 1°$$

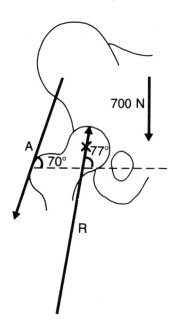

Figure 7.9.5. Forces acting on the hip joint during a one-leg stance (Trendelenburg test). The body weight excluding the weight of the standing leg (700 N) acts vertically downward. The abductor muscle force (A) acts at a 70° angle to the transverse plane. The joint reaction force (R) acts at a 77° angle to this plane. The center of rotation of the hip joint is marked with an X.

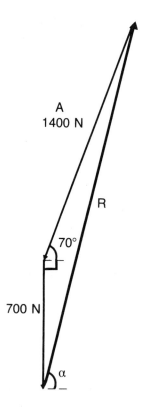

Figure 7.9.6. Polygon composed of the forces acting on the hip joint during a one-leg stance. The weight of the upper body (700 N) acts vertically downward, and the abductor muscle force (1400 N) acts at an angle of 70° to the transverse plane. The magnitude (R) and direction (α) of the hip joint reaction force are obtained.

Mathematical Solution

The hip joint reaction force R is obtained from Figure 7.9.6 by means of the cosine theorem.

$R^2 = 1400^2 + 700^2 - 2 \cdot 1400 \cdot 700 \cdot \cos 160°$

$R = 2070$ N

(For an alternate solution, by means of the Pythagorean theorem, see the mathematical solution to problem 7.7B.)

The direction (α) of the joint reaction force relative to the transverse plane is obtained by means of the sine theorem.

$$\frac{R}{\sin 160°} = \frac{1400}{\sin (90° - \alpha)}$$

where

$\beta = 90° - \alpha$

$$\frac{R}{\sin 160°} = \frac{1400}{\sin \beta}$$

$\beta = 13°$

$\alpha = 90° - 13°$

$\alpha = 77°$

7.5A. Find the external force (F) that must be applied at the ankle of a supine patient with the hip joint fully extended for the abductor muscle force to equal that in example 7.4.

Consider the leg as a free body. In this example, the clockwise moment is positive.

$\Sigma M = 0$ gives the magnitude of the external force F.

$70 - F \cdot 0.8 = 0$

$F = 88$ N

Find the magnitude and the direction α of the hip joint reaction force R relative to the transverse plane.

$\Sigma F = 0$ gives the joint reaction force R.

Graphic Solution

The magnitude and direction of force R are obtained by means of the polygon method (Figure 7.9.7).

$R = 1432 \pm 20$ N

$\alpha = 67 \pm 1°$

Mathematical Solution

The magnitude of the hip joint reaction force R is obtained from Figure 7.9.7 by means of the cosine theorem.

$R^2 = 1400^2 + 88^2 - 2 \cdot 1400 \cdot 88 \cdot \cos (180° - 70°)$

$R = 1432$ N

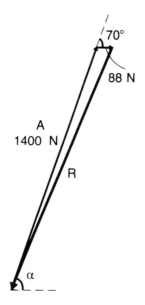

Figure 7.9.7. Polygon composed of the forces acting on the hip joint during abductor muscle exercises performed in the supine position. The external force (88 N) is applied parallel to the transverse plane. The abductor muscle force (1400 N) acts at a 70° angle to the transverse plane. The magnitude (R) and the direction α of the hip joint reaction force are obtained.

The direction (α) of force R relative to the transverse plane is obtained by means of the sine theorem.

$$\frac{1400}{\sin \alpha} = \frac{1432}{\sin 110°}$$

$$\alpha = 66.7°$$

7.6A. Find the external torque (M) on the hip joint when a cane is used in the hand on the side opposite the standing leg.

Consider the upper body as a free body. The force through the cane (150 N) acts vertically upward on the patient's hand and rotates the upper body counterclockwise. In this problem, the clockwise moment is positive.

$$M = 700 \cdot 0.1 - 150 \cdot 0.3$$

$$M = 25 \text{ Nm}$$

Find the joint reaction force (R) with a direction α relative to the transverse plane. The following forces act on the free body: the weight of the upper body (700 N), which acts vertically downward; the force through the cane (150 N), which acts vertically upward; the abductor muscle force (A), which acts at an angle of 70° to the transverse plane; and the joint reaction force (R).

Graphic Solution

The abductor muscle force (A) is calculated first, and the joint reaction force (R) is then determined. In this problem, the clockwise moment is positive.

ΣM = 0 gives the magnitude of the abductor muscle force A.

$25 - A \cdot 0.05 = 0$

$A = 500 \text{ N}$

$\Sigma F = 0$ gives the joint reaction force R.

The magnitude and direction of force R are obtained by means of the polygon method, as in Figure 7.9.6. (The force through the cane, directed vertically upward, is included.)

$R = 1050 \pm 25 \text{ N}$

$\alpha = 81 \pm 1°$

Mathematical Solution

The mathematical solution is obtained by means of the same method shown for problem 7.4B.

Answers to Discussion Questions for 7.4 through 7.6

1. The adducting torque on the hip joint produced by the weight of the upper body during the one-leg stance (Trendelenburg test) is 70 Nm for a person weighing 82 kg. The magnitude of this torque is determined by the weight of the upper body and its moment arm relative to the center of rotation of the hip joint.

2. If the ratio between the moment arms of the abductor muscle force and the body weight is high (i.e., the moment arm of the abductor muscle force is long, whereas the moment arm of the body weight is short), less abductor muscle force is required than if the ratio is low. A low muscle force means a low joint reaction force. This fact is of significance primarily in orthopaedic surgery, through which the moment arms of both the abductor muscle force and the body weight may be influenced to some extent. In arthroplasty, for example, the hip joint prosthesis can be placed medially to minimize the moment arm of the body weight and to maximize the moment arm of the abductor muscle force (Frankel and Nordin, 1980). Studies indicate that the ratio between the moment arms of the abductor muscle force and the body weight during a one-leg stance is 1:1–1:2 (Merchant, 1965; Morris, 1971).

3. For a torque of 70 Nm to be produced, an external force of 88 N must be applied at the ankle when the patient lies on his back, and 35 N when he lies on his side. Experience reveals that these figures may be high. It is not often that an external force of 80–90 N can be applied at the ankle of a patient lying on his back, even if the Trendelenburg test is negative. One explanation could be that the external moment arm of 0.1 m (Merchant, 1965; Morris, 1971), assumed in example 7.4, is too long, implying that during the Trendelenburg test the patient automatically shifts his weight toward the standing leg.

Another explanation may be that it is more physiologic, and therefore easier, to stand on one leg than to abduct the hip joint from a supine position. If this explanation is correct, then abductor muscle strengthening exercises should be performed as often as possible with the patient standing, provided that the magnitude of the joint reaction force does not need to be minimized.

4. With an adducting torque on the hip of 70 Nm, the joint reaction force (R) during a one-leg stance is about 2000 N, or 2.5 times body weight, whereas in the supine position it is 1400 N, or about 1.7 times body weight. The direction of the joint reaction force is also changed somewhat (i.e., from 77° to the transverse plane during a one-leg stance to 65° in the supine position). Therefore, the direction of the joint reaction force more closely parallels the long axis of the femoral neck when the patient is supine than when he stands on one leg.

During abductor muscle strengthening exercises in the reclining position, the hip joint reaction force is about the same whether the patient lies on his back or on his side. If the magnitude of the hip joint reaction force needs to be minimized, the patient should exercise in the reclining position rather than in the standing position. The hip joint reaction force is larger during standing because of the magnitude and direction of the weight of the upper body.

5. The example shows that cane use causes a greater reduction of the joint reaction force than does a weight loss of 15 kg. Denham (1959) and Morris (1971) showed that if the patient loses weight, the joint reaction force is reduced by an amount equal to three times the weight reduction. This value is in good agreement with the result in problem 7.6B, in which the patient loses 15 kg and the joint reaction force diminishes by 440 N.

If the patient uses a cane while standing on one leg (problem 7.6A), the joint reaction force decreases by nearly seven times the force exerted on the cane. (Transferring 150 N to the cane decreases the joint reaction force by about 1000 N.)

Blount (1956) demonstrated the value of the cane as a load-reducing aid. His calculations showed that the hip joint reaction force was reduced by about 1200 N when a force of 150 N was exerted on the cane; i.e., the reduction was equal to eight times the force on the cane. These values are correct provided that the patient uses the cane in a natural and effective way. Frankel (1979) showed that when the patient has difficulty using the cane naturally, the hip joint reaction force is the same whether or not a cane is used. By shifting the position of his upper body and changing his gait, the patient alters the location of his center of gravity in such a way that the external torque on the hip joint with cane use is the same as that without it.

Solution to Problem 7.7B

7.7B. Find the joint reaction force (R) on the right hip joint for a person standing on the right leg with a 25-kg suitcase in each hand.

Graphic Solution

Consider the upper body and the two suitcases as a free body. The following forces act on the free body: the weight of the upper body (500 N); the weight of the suitcases acting vertically downward (250 N + 250

N), the abductor muscle force (A), which acts at a 70° angle to the transverse plane; and the joint reaction force (R), whose line of application passes through the center of rotation of the hip joint.

The abductor muscle force (A) is calculated first, and then the joint reaction force (R).

In this problem, the clockwise moments are positive.

$\Sigma M = 0$ gives the magnitude of the abductor muscle force (A).

$$500 \cdot 0.06 + 250 \cdot 0.31 - 250 \cdot 0.14 - A \cdot 0.05 = 0$$

$$A = 1450 \text{ N}$$

$\Sigma F = 0$ gives the magnitude of the joint reaction force (R).

Force R is obtained by means of the polygon method, as in Figure 7.9.6. (The weight of the suitcases, totaling 500 N acting vertically downward, is included.)

$$R = 2410 \pm 30 \text{ N}$$

Mathematical Solution

The hip joint reaction force (R) is obtained by first calculating its vertical force component (R_y) and its horizontal force component (R_x) and then calculating the resultant (R) of these components.

In this problem, the forces directed vertically downward and horizontally to the left are positive.

$\Sigma F_y = 0$ gives the magnitude of the vertical force component R_y of the hip joint reaction force R.

$$500 + 250 + 250 + 1450 \sin 70° - R_y = 0$$

$$R_y = 2362 \text{ N}$$

$\Sigma F_x = 0$ gives the magnitude of the horizontal force component R_x of the hip joint reaction force R.

$$1450 \cos 70° - R_x = 0$$

$$R_x = 496 \text{ N}$$

The resultant (R) of the vertical and horizontal force components is obtained by means of the Pythagorean theorem.

$$R^2 = 2362^2 + 496^2$$

$$R = \sqrt{5825060}$$

$$R = 2414 \text{ N}$$

For an alternative solution by means of the cosine theorem, see the mathematical solution to problem 7.4B.

Answer to Discussion Question in 7.7

In this example, the hip joint reaction force is larger when 25 kg is carried in one hand than when 50 kg is distributed evenly between two hands. Distribution of the burden affects the magnitude of the joint reaction force so greatly that it can be considered an important point to be brought up during ergonomic counseling. An equally important point is that the object should be carried close to the body.

Solutions to Problems 7.8A and B

7.8A. Find the bending moment (M) in the frontal plane acting at an arbitrarily selected center of rotation in a total hip joint prosthesis secured to the bone in a varus position.

Consider the part of the prosthesis above the center of rotation as a free body (Figure 7.9.8, left). The joint reaction force R and the supporting force C in the calcar region act on the free body. In this problem, the counterclockwise moment is positive.

$$M = 2800 \cdot 0.04 - 1400 \cdot 0.02$$

$$M = 84 \text{ Nm}$$

7.8B. Find the bending moment (M) in the frontal plane at the center of rotation in a total hip joint prosthesis secured in a valgus position

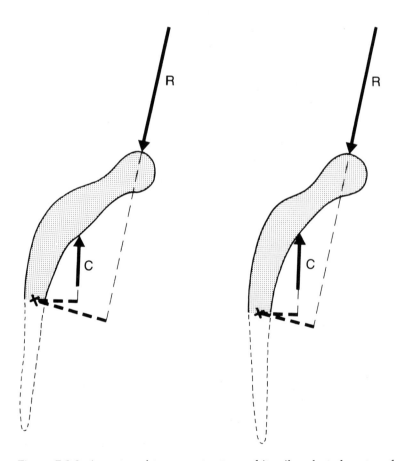

Figure 7.9.8. An external torque acts at an arbitrarily selected center of rotation (X) in a total hip joint prosthesis secured in a varus position (left), and in a valgus position (right). The joint reaction force (R) and the supporting force (C) in the calcar region are shown, along with their moment arms (broken lines).

(Figure 7.9.8, right). In this problem, the clockwise moment is positive.

$$M = 2800 \cdot 0.03 - 1400 \cdot 0.02$$

$$M = 56 \text{ Nm}$$

Answers to Discussion Questions in 7.8

1. When the load is transmitted through the head of the total hip joint prosthesis, a bending moment in the frontal plane, and hence tensile stress, develop on the lateral side of the prosthesis. Valgus positioning of the prosthesis, as opposed to varus positioning, reduces the bending moment and consequently the tensile stress on the lateral side. As the prosthesis comes under repeated loading, this reduction in tensile stress can significantly lengthen the life of the prosthesis. Thus, valgus positioning of the prosthesis is an important factor in combating fatigue failure of the prosthesis.

2. Another factor affecting the bending moment in the frontal plane is insufficient support for the prosthesis in the calcar region, either because the prosthesis is not entirely surrounded by cement, because it loosens in the cement, or because the bone in the calcar region becomes necrotic. The prosthesis then "wiggles" under the load, and tensile stress is produced on the lateral side. The significance of such factors has been detailed by Andriacchi et al. (1976), who pointed out that adequate support for the prosthesis in the calcar region, as well as valgus positioning of the prosthesis, is of great importance in reducing the bending moment.

Still another factor affecting the bending moment is the design of a prosthesis. A design that creates a long moment arm, such as a sharply angled neck, increases the bending moment. The mechanical properties and the dimensions of a prosthesis are also important, since they affect the magnitude of the bending moment that the prosthesis can sustain before failure.

7.10. Commentary

The Torque Counteracted by the Abductor Muscles during Walking and during Muscle Strength Testing with a Dynamometer

It is interesting to compare the torque on the hip joint produced by the hip abductor muscles during muscle strengthening exercises with the torque produced during normal walking. Andriacchi et al. (1981) studied walking in 29 healthy individuals (15 women and 14 men; age range, 21 to 60 years). The pattern of hip motion studied in the frontal plane was uniform among the subjects and was independent of sex and walking speed. In the stance phase, an adducting torque placed demands on the abductor muscles. The adducting torque remained approximately the same throughout the entire stance phase, amounting to 5% of the individual's body weight multiplied by the height of the body. For an individual with a weight of 82 kg and a height of 1.71 m, the adducting torque on the hip joint during walking is therefore $0.05 \cdot 820 \cdot 1.71 = 70$ Nm.

The maximum isometric torque counteracted by the hip abductor muscles of 80 healthy persons (40 women and 40 men) was studied by Murray and Sepic (1968). Half of the subjects were between 40 and 55 years of age, and half were between 18 and 33 years. During strength tests of the abductor muscles with the subjects in a supine position, the older women achieved a torque of 73 Nm, while the younger ones achieved 89 Nm. The maximum isometric torque for the older men was 120 Nm and for the younger ones, 136 Nm.

A comparison of these two studies shows that the adducting torque during walking amounts to 50% or more of the maximum force that the abductor muscle group can counteract isometrically. Thus, walking is an excellent exercise for the hip abductor muscles. For such a comparison to be valid, the same parts of the abductor musculature must be contracted and the length-tension relationship of the muscles being tested must be the same in both situations. The hip joint angle in all three planes (sagittal, frontal, and transverse), as well as the muscle contraction time, must also be standardized for both situations.

In the symmetrical standing position, the line of gravity passes posterior to the pubic symphysis, and since the hip joints are stable, the muscle activity around these joints is low (Carlsöö, 1972). If the leg weighs one-sixth of body weight, the joint reaction force in each hip joint will be one-third of body weight (Frankel and Nordin, 1980; Rydell, 1966). The hip joint reaction force was calculated to be two to five times body weight during walking (Paul and McGrouther, 1975; Seireg and Arvikar, 1975; Crowninshield et al., 1978; Johnston et al., 1979). During very fast walking (Paul and McGrouther, 1975) and stair climbing (Johnston et al., 1979), values were as high as seven times body weight.

Intravital measurements using instrumented prostheses in two patients (Rydell, 1966) showed that hip joint reaction forces varied between 1.6 and 3.3 times body weight during activities of daily living.

The values appearing elsewhere in the literature are the result of biomechanical calculations in which certain assumptions were made in solving the problems. In dynamic two- and three-dimensional calculations of joint reaction forces during walking, analytical methods and devices such as films, force plates, and electromyography are used. Electromyography is used directly in the analysis of forces during walking to identify the muscle activity producing the forces (Crowninshield et al., 1978) or as a means of verifying muscle activity determined by other methods (Seireg and Arvikar, 1975). The external torque is determined from the films and the force plates, and the joint reaction force is then calculated. In the literature, two different mathematical models are often used for this calculation. In one model, the reduction method, the least significant forces are disregarded and the number of unknowns is reduced to six, matching the number of equilibrium equations (three moment equilibrium equations and three force equilibrium equations, two in each coordinate direction).

The other model, linear programming, is an optimizing method in which there are more unknown variables than equations. Adding a number of boundary conditions allows a group of possible equation solutions to be constructed. The best of these (in some respect) is then selected (see also Chapter 9, Section 9.5, Commentary).

Direction of the Joint Reaction Force

During the one-leg stance, the hip joint reaction force acts at an angle of 75–85° to the horizontal plane (Inman, 1947). Its direction is determined by the directions of the muscle force and the weight of the upper body.

The direction of the joint reaction force parallels the longitudinal orientation of the medial trabeculae, thereby enhancing the strength of the femoral neck (Frankel, 1960). In other words, the abductor muscles help to create a mechanically advantageous direction for the joint reaction force.

Assumptions Made in Examples 7.3 through 7.7

1. It is assumed that the abductor muscles absorb the entire adducting torque in the hip joint. This assumption disregards any forces produced by the ligaments, joint capsule, or other soft tissues. In a study of the function of patients in whom one or more muscles in the lower extremities had been removed surgically because of soft tissue tumors, it was found that all seven patients lacking one or more abductor muscles had a positive Trendelenburg test (Markhede, 1980). The validity of the Trendelenburg test was also studied in 90 patients who had undergone total hip joint replacement (Johnston and Larsson, 1969). A positive Trendelenburg test was seen less frequently in patients with the least shortening of the distance between the origin and the point of attachment of the abductor muscles and the longest muscle force moment arm. Both of these studies attest to the plausibility of the assumption.

2. The moment arm of the abductor muscle force is assumed to be 5 cm. Within the range of hip motion from 10° of adduction to 10° of abduction, the moment arm of the abductor muscle force varies between 4.4 and 4.8 cm, depending on the hip joint angle (Olson et al., 1972). With allowances for uncertainties in the method of measurement, 5 cm is a reasonable assumption.

3. The line of application of the abductor muscle force is assumed to form a 70° angle with a horizontal line through the trochanters (transverse plane). This line of application has been calculated to act at a 71° angle

to this plane (Williams and Lissner, 1977) and is mainly dependent on the magnitudes and directions of the forces produced by each individual abductor muscle.

4. The moment arm of the body weight during a one-leg stance is assumed to vary from 6 to 10 cm in examples 7.4 through 7.8. The ratio between the moment arms of the abductor muscle force and the body weight is cited in the literature as 1:1–1:2 (Merchant, 1965; Morris, 1971). If the moment arm of the abductor muscle force is assumed to be 5 cm, the moment arm of the body weight can therefore be assumed to be between 5 and 10 cm.

5. In example 7.6, it is assumed that a load of 150 N can be transferred to a cane. The average maximum force transferred to a cane during one gait cycle was 180 N in a study of 20 patients with joint disorders (Murray et al., 1969).

8

The Knee

The knee is composed of the tibiofemoral and the patellofemoral joints. During flexion-extension of the tibiofemoral joint, motion takes place in all planes, but the range of motion is substantially greater in the sagittal plane than in the other two planes. The axis of motion in the sagittal plane passes through the femoral epicondyles (Frankel and Nordin, 1980). This axis shifts only slightly throughout the range of knee joint motion and thus the joint can be considered a hinge.

All of the examples in this chapter study knee joint motion in the sagittal plane and involve the forces produced by the quadriceps muscle. Five examples involve calculations of the torque on the knee joint (8.1–8.5); in one of these, the magnitude of the force produced by the hamstring muscles is also calculated (8.3). Example 8.6 demonstrates the composition of forces acting on the patellofemoral joint. Three examples (8.7–8.9) demonstrate the two conditions for static equilibrium—force equilibrium and moment equilibrium. Two of these examples (8.7 and 8.8) compare the loads on the tibiofemoral joint during various quadriceps muscle exercises. In the last example (8.9), the joint reaction force during stair climbing is compared with that during stair descending for both the tibiofemoral and the patellofemoral joints.

8.1. Change in the Direction of an Externally Applied Force during Quadriceps Muscle Exercises

Torque

This example can be used to review the effect of the moment arm of a force on the torque produced. The example shows that if the point of application of a force remains constant, a change in the direction affects the moment arm, and thus the torque.

Problems

The isometric strength of the knee extensor muscles is tested in a seated patient. With the knee flexed 30°, a torque of 210 Nm is produced. This value for the torque is taken from a study of the knee extensor muscle strength of 10 healthy young men (Scudder, 1980).

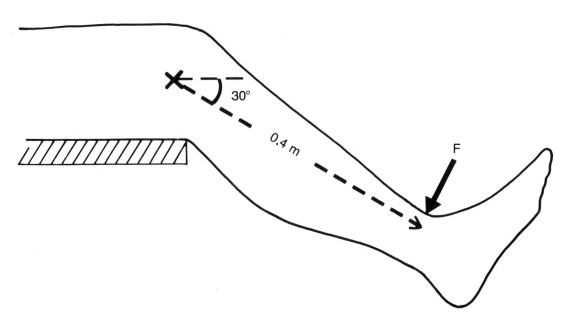

Figure 8.1A. Quadriceps muscle exercise with the patient seated and the knee flexed 30°. An external force (F) is applied perpendicular to the long axis of the tibia 0.4 m distal to the center of motion of the knee joint.

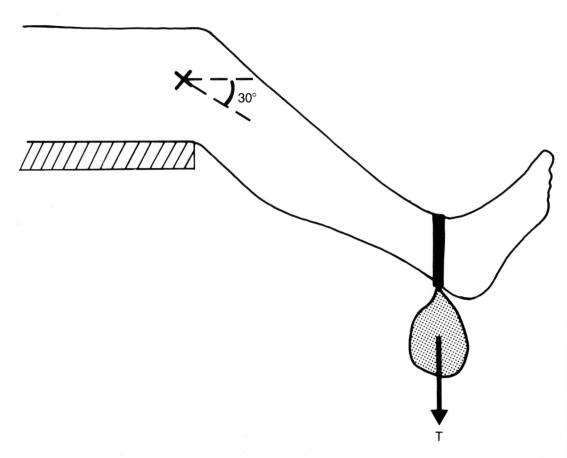

Figure 8.1B. A sandbag with a weight T is hung from the ankle during the same exercise shown in Figure 8.1A.

8.1A. What is the maximum force (F) that can be applied perpendicular to the long axis of the tibia 0.4 m distal to the center of motion of the knee joint before the knee flexes (Figure 8.1A)? Disregard the weight of the lower leg.

8.1B. In a variation of the above situation, a sandbag is hung at the point at which the external force (F) was applied (Figure 8.1B). How heavy a sandbag can be hung before the knee flexes? The weight produced by the sandbag is designated as T. Disregard the weight of the lower leg.

8.2. The Step Test. A Functional Strength Test of the Knee Muscles

Torque

In this example, the body weight acts as an external force during the functional strength test of stepping up onto a bench. The starting point for the calculation of the torque is the point at which the flexing torque on the knee joint is maximal. This point is reached when the rear foot is no longer in contact with the floor. In the positions shown in Figure 8.2, the patient has stopped moving (i.e., static equilibrium has been achieved).

Problems

The strength of a patient's knee muscles during extension is examined by means of the step test. The patient weighs 64 kg, and the lower leg weighs 4 kg.

8.2A. How large a torque (M) must the knee extensor muscles counteract when the patient steps up onto a low bench (Figure 8.2A)? The moment arm for the body weight (T) acting above the right knee is 0.17 m.

8.2B. How large a torque (M) must the knee extensor muscles counteract when the patient steps up onto a high bench (Figure 8.2B)? The moment arm for the body weight (T) acting above the right knee is 0.18 m.

Discussion Question

Why does the patient encounter greater difficulty in stepping up onto the high bench than onto the low bench, even though the torque on the knee joint is approximately the same in both cases?

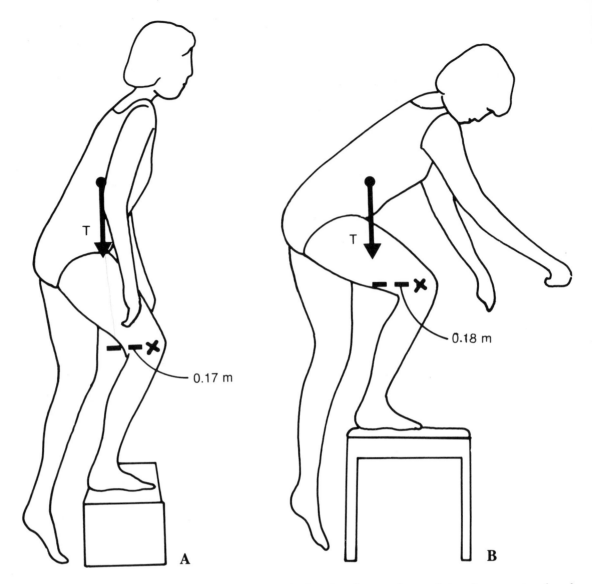

Figure 8.2. The patient performs the functional strength test of stepping up onto a bench (step test). A. When a low bench is used, the moment arm of the weight of the body above the right knee (T) is 0.17 m. B. When a high bench is used, the moment arm of T is 0.18 m.

8.3. Strength Testing of the Knee Flexor and Extensor Muscles with the Dynamometer

Torque

This example illustrates the relationship between external and internal torque.

Problems

The isometric strength in the extensor and flexor muscles of the knee is measured at four different angles of the knee joint. The patient is lying on his side with the hip extended and the thigh restrained. A dyna-

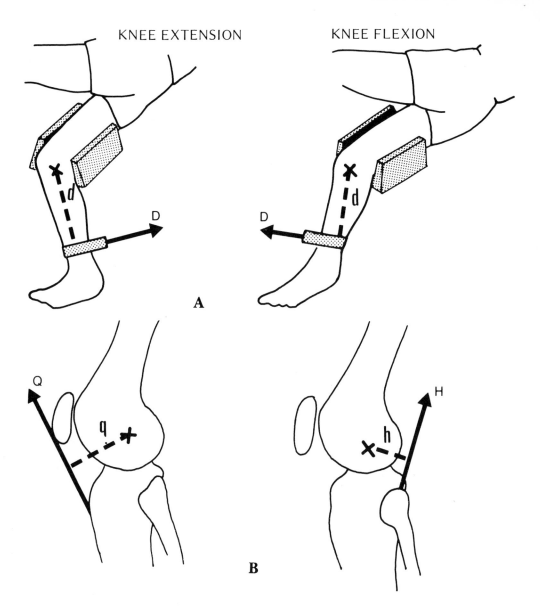

KNEE EXTENSION   KNEE FLEXION

A

B

Figure 8.3. Strength testing of the extensor and flexor muscles of the knee. A. The external torque on the knee is the product of the dynamometer force (D) and its moment arm (d). B. The internal torque on the knee is the product of the quadriceps muscle force (Q) and its moment arm (q) (left), and the product of the hamstring muscle force (H) and its moment arm (h) (right). (From Smidt, 1973.)

mometer is attached to the ankle perpendicular to the long axis of the tibia (Figure 8.3A).

The table below gives the torques obtained and the moment arm lengths calculated for the quadriceps muscle force and the hamstring muscle force. The values are based on a study of 26 healthy men ranging in age from 19 to 47 years (Smidt, 1973).

In the problems, assume that the quadriceps muscle is the only active extensor muscle in the knee joint, and that the hamstring muscles are the only active flexor muscles.

Knee Joint Angle (°)	Extending Torque (Nm)	Quadriceps Muscle Force Moment Arm (cm)	Flexing Torque (Nm)	Hamstring Muscle Force Moment Arm (cm)
5	61	4.4	64	2.5
45	121	4.9	50	4.1
60	120	4.7	51	3.9
90	104	3.8	36	2.6

8.3A. What values for force appear on the dynamometer (D) when it is attached 0.32 m distal to the center of motion of the knee joint?

8.3B. How large a quadriceps muscle force (Q) or hamstring muscle force (H) is required at the various knee joint angles to counteract the external torque (Figure 8.3B)?

Discussion Question

What information must be recorded for measurements of knee muscle strength to be comparable in repeated strength tests with a dynamometer?

8.4. Quadriceps Muscle Exercises with a Weight Boot

Torque

The torque on the knee joint produced by a weight boot during quadriceps muscle exercises performed in a sitting position is compared with the maximal isometric torque against resistance produced by the quadriceps muscle during knee motion from 90° of flexion to full extension. The values for maximal torque were obtained from a study of 6 healthy men ranging in age from 20 to 40 years (Lindahl et al., 1969).

Problems

8.4A. From a sitting position, a patient is exercising the knee extensor muscles with a 10-kg weight boot. How large is the external torque (M) when the patient keeps his leg steady at 90°, 60°, 30°, and 15° of knee flexion (Figure 8.4)? The lower leg weighs 5 kg, and its center of gravity is located 0.2 m from the center of motion of the knee joint. The distance from the center of motion to the center of gravity of the weight boot is 0.45 m. Using these four values to calculate the torque, construct a rectangular coordinate system in which the horizontal axis represents the angle of knee joint flexion and the vertical axis represents the magnitude of the torque.

8.4B. The maximum isometric strength in the knee extensor muscle for the same patient is as follows:

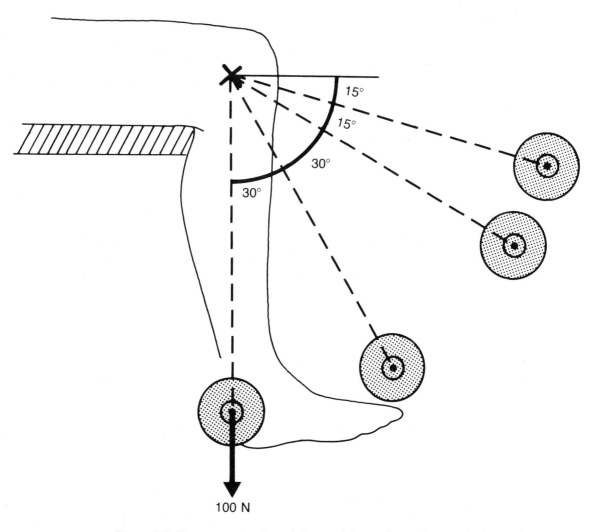

Figure 8.4. Knee extension in a sitting position with a 10-kg weight boot. The position of the weight boot relative to the center of motion of the knee is shown for four angles of the knee joint (90°, 60°, 30°, and 15°). (From Williams and Lissner, 1977.)

Knee Joint Angle (°)	Maximal Isometric Extending Torque (Nm)
90	175
60	230
30	140
15	90

Plot these torque values on the coordinate system constructed in problem 8.4A.

Discussion Questions

1. What percentage of the maximal torque is required for the knee extensor muscle to resist the force produced by the weight boot at 60°

and at 15° of knee flexion? At what joint angle will the exercise be most effective?

2. Will the patient be able to extend the knee joint the last 15° while wearing the weight boot? In discussing this question, make use of your experience with quadriceps muscle exercises.

8.5. Exercising the Knee Extensor Muscles in a Sitting Position on the Quadriceps Table

Torque

This example shows that if the magnitude and direction of a force are held constant, a change in the point of application affects the moment arm, and thus the torque.

Problems

A patient is exercising the knee extensor muscles while sitting on the quadriceps table. A 10-kg weight is placed on the weight arm 0.35 m from the center of motion of the knee joint.

8.5A. How large a torque must the knee extensor muscles counteract at full extension (M_1) (Figure 8.5A) and at 60° of knee flexion (M_2) (Figure 8.5B) if the weight arm and the resistance arm are parallel? Disregard the weight of the lower leg and the weight of the weight arm and resistance arm.

8.5B. Because the knee extensor muscles can normally produce less torque at full extension than at 60° of knee flexion, the weight arm is moved upward at a 60° angle from the resistance arm. With this placement of the weight arm, the external torque reaches its maximum at 60° of knee flexion; the knee extensor muscle force is also usually the largest at this point in the range of motion. With the knee joint in full extension, how heavy a weight must be attached to the weight arm for the torque to equal that in problem 8.5A (Figure 8.5C)? The force produced by the attached weight is designated as T.

Figure 8.5. Knee extensor muscle exercises with the individual in a sitting position on the quadriceps table. With the weight arm and resistance arm parallel, the knee joint is fully extended (Figure A) and flexed 60° (Figure B). In Figure C, the weight arm is moved upward at a 60° angle to the resistance arm, and the knee joint is in full extension.

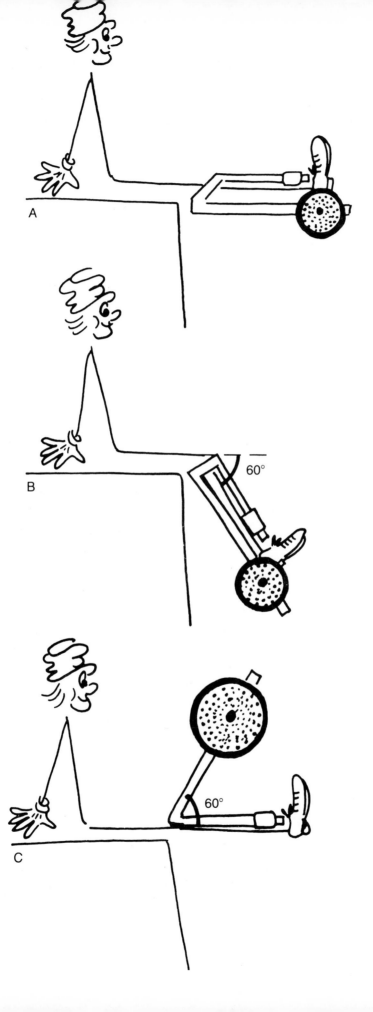

A

B

60°

C

60°

8.6. The Patellofemoral Joint Reaction Force during Quadriceps Muscle Exercises with the Knee Extended and Flexed

Composition of Forces

This example shows the effect of the knee joint angle on the patellofemoral joint reaction force during quadriceps muscle exercises. It can be used in planning and discussing exercise programs for patients with chondromalacia patellae, in whom the magnitude of the patellofemoral joint reaction force must be taken into consideration.

The angle between the patellar tendon and the quadriceps tendon (Figure 8.6) varies with the knee joint angle. In the example, the values for this angle are taken from a study by Matthews et al. (1977), who determined the angle radiographically after placing two metal wires along each of these tendons.

Problems

8.6A. A patient is exercising the quadriceps muscle isometrically from a sitting position with the knee flexed 5°. The quadriceps muscle force (Q) is 1000 N. The angle between the patellar tendon and the quadriceps tendon is 35° (Figure 8.6, top). How great is the patellofemoral joint reaction force (P) at equilibrium? The patella can be considered as a frictionless pulley. The force transmitted through the patellar tendon is designated as T.

8.6B. The patient exercises the quadriceps muscle with the knee flexed 45°, and then 90°. The quadriceps muscle force in both cases is 1000 N.

How large is the patellofemoral joint reaction force (P) at the two joint angles if the angle between the patellar tendon and the quadriceps tendon is 55° at 45° of knee flexion (Figure 8.6, middle) and 80° at 90° of knee flexion (Figure 8.6, bottom)? Consider the patella as a frictionless pulley.

Plot the values for forces Q and P for each of the three joint angles in a coordinate system in which the horizontal axis represents the knee joint angle and the vertical axis represents the magnitude of the forces.

Discussion Question

Cite some factors that influence the magnitude of the patellofemoral joint reaction force.

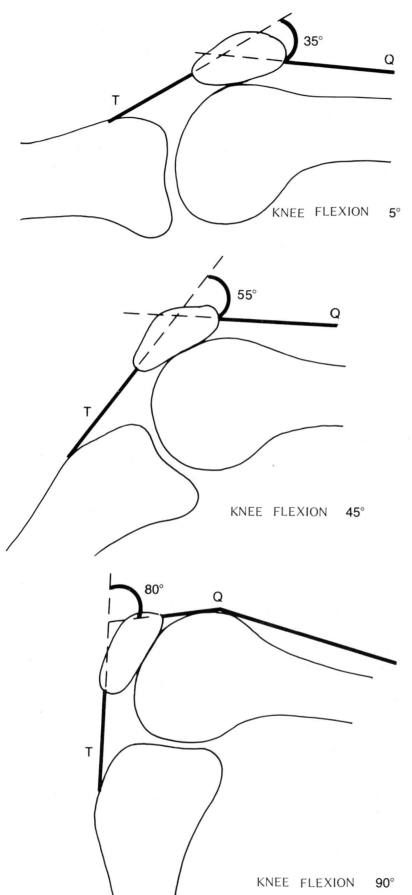

35°

Q

T

KNEE FLEXION   5°

55°

Q

T

KNEE FLEXION   45°

80°   Q

T

KNEE FLEXION   90°

Figure 8.6. The angle between the patellar tendon (T) and the quadriceps tendon (Q) is 35° at 5° of knee flexion (top), 55° at 45° of knee flexion (middle), and 80° with the knee flexed 90° (bottom).

8.7. Loading of the Tibiofemoral Joint during Quadriceps Muscle Exercises in the Standing and Sitting Positions

Moment Equilibrium; Force Equilibrium

This example allows a comparison of the tibiofemoral joint reaction force during quadriceps muscle strengthening exercises in the sitting and standing positions with the quadriceps muscle force held constant.

The value for the quadriceps muscle force during exercises in the standing position is based on the results of a study of 3 women and 9 men (age range, 18 to 38 years) in which the forces on the tibiofemoral joint during the gait cycle were calculated (Morrison, 1970). The maximum quadriceps muscle force during the stance phase averaged 740 N and occurred when the knee was flexed about 15° after heel strike and then extended again.

Problems

A patient weighing 70 kg is standing on one leg with the knee of the standing leg flexed 15°. The quadriceps muscle force required to maintain this position is 740 N.

8.7A. The patient is seated with the knee flexed 15° (Figure 8.7A). A vertical external force (F) is applied to the ankle 0.4 m from the center of motion of the knee joint. How large must the external force (F) be for the quadriceps muscle force to equal that during the one-leg stance? Disregard the weight of the lower leg. The moment arm of the quadriceps muscle force is 0.05 m.

8.7B. How large is the tibiofemoral joint reaction force (R) when the patient performs the exercise in the standing and in the sitting positions? Disregard the weight of the lower leg. With the knee flexed 15°, the line of application of the patellar tendon force forms a 20° angle with the long axis of the tibia. With the patient in the standing position, the long axis of the tibia forms an 80° angle with the horizontal plane (Figure 8.7B).

Draw in the forces acting on the lower leg in Figures 8.7A and B.

Discussion Question

State some advantages of performing quadriceps muscle exercises in the standing position and in the sitting position.

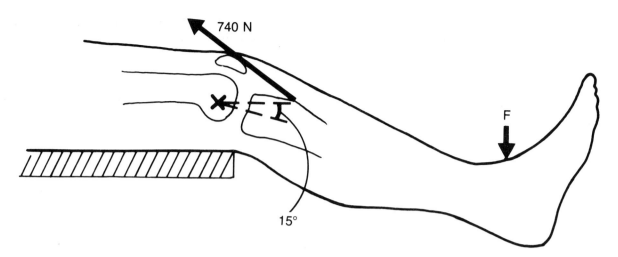

Figure 8.7A. Quadriceps muscle exercise with the patient in a sitting position and the knee flexed 15°. A vertical external force (F) applied to the ankle produces a quadriceps muscle force of 740 N.

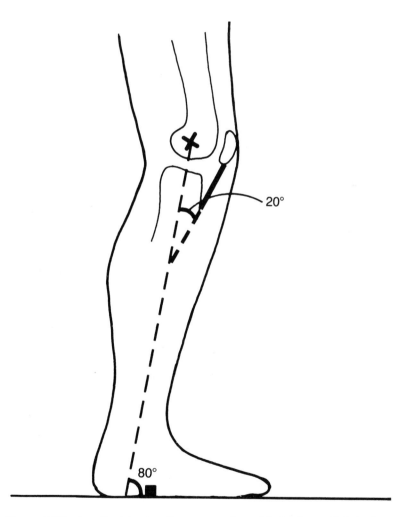

Figure 8.7B. A patient is standing on one leg with the knee slightly flexed. The patellar tendon force forms a 20° angle with the long axis of the tibia, and the tibia forms an 80° angle with the horizontal plane. The point of application of the ground reaction force (equal to the body weight) is marked with a solid square.

8.8. Loading of the Tibiofemoral Joint during Quadriceps Muscle Exercises with the Knee Extended and Flexed

Moment Equilibrium; Force Equilibrium

The quadriceps muscle is exercised isometrically with the knee flexed 30° and 90°. Because the purpose of the example is to show the effect of the knee joint angle on the internal forces on the tibiofemoral joint, the external force is kept constant.

The angles given for the line of application of the patellar tendon force are based on the average angles for 26 young men (Smidt, 1973).

Problems

8.8A. A patient is sitting and exercising the quadriceps muscle with the knee flexed 30°. At the ankle, 0.4 m from the center of motion of the knee, an external force of 150 N is applied perpendicular to the long axis of the tibia (Figure 8.8A). The moment arm of the quadriceps muscle force is 0.05 m. The force transmitted through the quadriceps tendon acts at an angle of 20° to the long axis of the tibia. The tibial plateau is perpendicular to this long axis. The weight of the lower leg may be disregarded.

How large a force (Q) must the quadriceps muscle exert for the lower leg to remain in 30° of flexion?
How large is the tibiofemoral joint reaction force (R)?
How large is the shear force (S) on the tibiofemoral joint?
In Figure 8.8A, draw in the force components acting on the lower leg parallel to the tibiofemoral joint surface.

8.8B. The same patient flexes the knee 90°. The patellar tendon then parallels the long axis of the tibia (Figure 8.8B). All other conditions are identical to those in problem 8.8A.

How large a force (Q) must the quadriceps muscle produce for the joint angle of 90° to be maintained?
How large is the tibiofemoral joint reaction force (R)?
How large is the shear force (S) on the tibiofemoral joint?
In Figure 8.8B, draw in the force components acting on the lower leg parallel to the tibiofemoral joint surface.

Discussion Questions

1. Is the magnitude of the tibiofemoral joint reaction force (R) affected by the knee joint angle?
2. Which cruciate ligament transmits loads during the quadriceps muscle exercise against resistance with the knee flexed 30°? With the knee flexed 90°? Assume that the anterior cruciate ligament absorbs ventrally directed shear forces and that the posterior cruciate ligament absorbs dorsally directed shear forces.

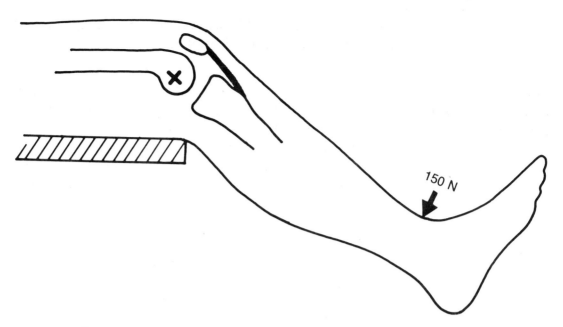

Figure 8.8A. Quadriceps muscle exercise with the patient in a sitting position and the knee flexed 30°. An external force of 150 N is applied to the ankle perpendicular to the long axis of the tibia. The alignment of the patellar tendon is shown.

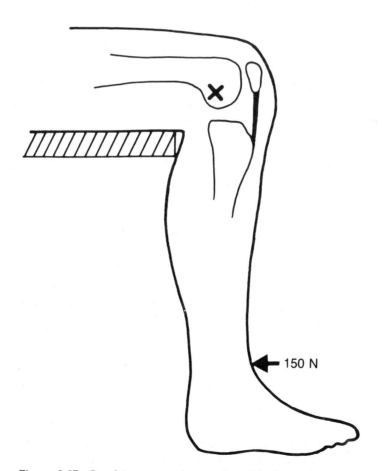

Figure 8.8B. Quadriceps muscle exercise with the patient in a sitting position and the knee flexed 90°. An external force of 150 N is applied to the ankle perpendicular to the long axis of the tibia. The alignment of the patellar tendon is shown.

8.9. Tibiofemoral and Patellofemoral Joint Reaction Forces during Stair Climbing and Descending

Moment Equilibrium; Force Equilibrium

This example allows a comparison of the tibiofemoral and patellofemoral joint reaction forces during stair climbing and descending at a slow, even pace.

The magnitude of the torque during stair climbing is based on a study of 10 men (average weight, 71 kg; average height, 1.79 m) (Andriacchi et al., 1980). The line of application of the patellar tendon force was obtained from Smidt (1973).

Problems

A patient is exercising the quadriceps muscles by using the stairs. The maximum external torque counteracted by the knee extensor muscles is 55 Nm during stair climbing and 135 Nm during stair descending.

8.9A. How large is the tibiofemoral joint reaction force (R) in the two cases? The quadriceps muscle is assumed to be the only active muscle, and it absorbs the entire flexing torque. The quadriceps muscle force transmitted through the patellar tendon (T) has a moment arm of 0.05 m.

During stair climbing, the force transmitted through the patellar tendon acts at an angle of 15° to the long axis of the tibia, and the tibia forms a 60° angle with the horizontal plane (Figure 8.9A). During stair descending, the force transmitted through the patellar tendon acts at an angle of 10° to the long axis of the tibia, and the tibia forms a 50° angle with the horizontal plane (Figure 8.9B). The ground reaction force is 700 N and is directed vertically upward during both stair climbing and descending. Disregard the weight of the lower leg.

8.9B. How large is the patellofemoral joint reaction force (P) during stair climbing and descending? The force transmitted through the quadriceps tendon is assumed to act parallel to the long axis of the femur. The force transmitted through the patellar tendon (T) acts at an angle of 15° to the long axis of the tibia during stair climbing, and at an angle of 10° during stair descending. The patella acts as a frictionless pulley. The angle between the long axes of the femur and the tibia is 130° during stair climbing (Figure 8.9A) and 115° during stair descending (Figure 8.9B). Obtain the necessary data from problem 8.9A.

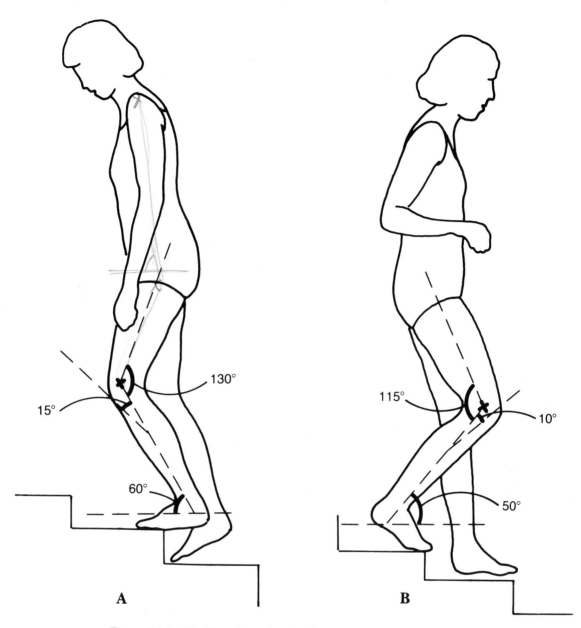

A

B

Figure 8.9A. Climbing the stairs slowly. The tibia forms a 60° angle with the horizontal plane. The force transmitted through the patellar tendon acts at a 15° angle to the long axis of the tibia. The long axes of the femur and tibia form a 130° angle with respect to each other.

Figure 8.9B. Descending the stairs slowly. The tibia forms a 50° angle with the horizontal plane. The force transmitted through the patellar tendon acts at a 10° angle to the long axis of the tibia. The long axes of the femur and tibia form a 115° angle with respect to each other.

8.10. Answers, Solutions, and Discussion

Answers to the Problems

8.1A. F = 525 N

8.1B. T = 606 ± 20 N

8.2A. 102 Nm

8.2B. 108 Nm

8.3A.

Knee Joint Angle	D (extension)	D (flexion)
5°	191 N	200 N
45°	378 N	156 N
60°	375 N	159 N
90°	325 N	112 N

8.3B.

Knee Joint Angle	Q	H
5°	1386 N	2560 N
45°	2469 N	1220 N
60°	2553 N	1308 N
90°	2737 N	1385 N

8.4A.

Knee Joint Angle	M
90°	0 Nm
60°	28 ± 2 Nm
30°	48 ± 2 Nm
15°	53 ± 2 Nm

8.5A. 0°: $M_1 = 35$ Nm
       60°: $M_2 = 18 ± 1$ Nm

8.5B. T = 194 ± 10 N

8.6.

Knee Joint Angle	P
5°	601 ± 20 N
45°	923 ± 20 N
90°	1285 ± 20 N

8.7A. F = 96 ± 3 N

8.7B. R (standing) = 1391 ± 20 N
      R (sitting) = 690 ± 20 N

8.8A. Q = 1200 N
      R = 1157 ± 20 N
      S = 260 ± 30 N

8.8B. Q = 1200 N
      R = 1209 ± 20 N
      S = 150 N

8.9A. R (climbing) = 1670 ± 20 N
      R (descending) = 3195 ± 40 N

8.9B. P (climbing) = 1182 ± 30 N

P (descending) = 3287 ± 30 N

The ranges for some of these answers allow for a reasonable margin of error in the graphic solution of the problems. The ranges were determined on the basis that the maximum allowable error in measurement is 1 mm of length and 1 degree of angle. In most examples, the figures were scaled so that 5 cm in reality corresponds to 1 cm in the figure and so that 100 N corresponds to 1 cm. In example 8.9, 200 N corresponds to 1 cm; and in example 8.4, 10 N corresponds to 1 cm.

Solutions to Problems 8.1A and B

8.1A. Find the external force (F) applied to the ankle during a test of the isometric strength of the knee extensor muscles.

The external force (F) is applied perpendicular to the long axis of the tibia, and its distance from the center of motion of the knee constitutes its moment arm. In this problem, the clockwise moment is positive.

$\Sigma M = 0$ gives the magnitude of force F.

$F \cdot 0.4 = 210$

$F = \dfrac{210}{0.4}$

$F = 525$ N

8.1B. Find the magnitude of the weight (T) produced by the sandbag hung at the point on the ankle at which the external force was applied in problem 8.1A.

Graphic Solution

The weight produced by the sandbag (T) acts vertically. The moment arm (a) is therefore the horizontal distance between the point of application of weight T on the proximal ankle and the center of motion of the knee joint. The moment arm (a) forms the side of a right triangle with a hypotenuse (the long axis of the tibia) of 0.4 m and one acute angle of 30° (Figure 8.10.1).

$a = 0.35 \pm 0.01$ m

$\Sigma M = 0$ gives the weight T.

$T \cdot 0.35 = 210$

$T = \dfrac{210}{0.35}$

$T = 606 \pm 20$ N

105

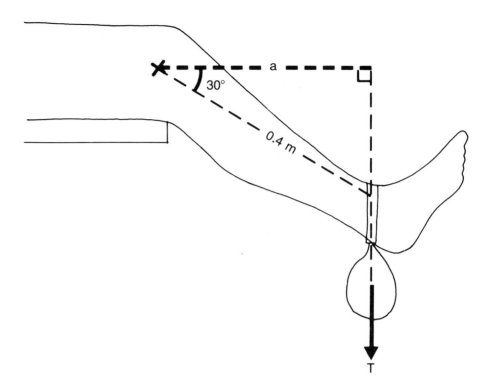

Figure 8.10.1. Quadriceps muscle exercise in a sitting position with the knee flexed 30° and a sandbag hung at the ankle, producing a weight (T) acting 0.4 m from the center of motion of the knee joint. The moment arm (a) of weight T relative to the center of motion of the knee joint forms the side of a right triangle, in which the hypotenuse is 0.4 m and one acute angle is 30°.

Mathematical Solution

The length of the moment arm (a) for the weight produced by the sandbag (T) is obtained from Figure 8.10.1.

$a = 0.4 \cos 30°$

$$T = \frac{210}{0.4 \cos 30°}$$

$T = 606$ N

Solution to Problem 8.2A

8.2A. Find the external torque (M) on the knee joint when the patient steps up onto a low bench. Weight T is equal to the body weight (640 N) minus the weight of the lower leg (40 N).

$T = 640 - 40$

$T = 600$ N

$M = 600 \cdot 0.17$

$M = 102$ Nm

106

Answer to Discussion Question in 8.2

The following are some explanations why stepping up onto a high bench is more difficult than stepping onto a low bench, even though the torque on the knee joint produced by the body weight is approximately the same in both cases:

- More lifting work is performed (for a definition of lifting work, see Chapter 9, example 9.3).
- The greater flexion of the knee joint (see Figure 8.2B) affects the maximum internal torque, requiring a greater percentage of the maximum strength of the quadriceps muscle.
- The hip extensor muscles must counteract a larger torque, since the upper body tilts farther forward (compare Figures 8.2A and B).

Solutions to Problems 8.3A and B

8.3A. Find the dynamometer force (D) produced during isometric strength tests of the knee extensor and flexor muscles.

The solution is given only for the dynamometer force produced during a knee extensor muscle strength test with the knee flexed 5°.

$$D = \frac{61}{0.32}$$

$$D = 191 \text{ N}$$

8.3B. Find the quadriceps muscle force (Q) and the hamstring muscle force (H) required to counteract the external torque produced by the dynamometer at various knee joint angles.

The solution is given only for the quadriceps muscle force (Q) at 5° of knee flexion.

$$Q = \frac{61}{0.044}$$

$$Q = 1386 \text{ N}$$

Answer to Discussion Question in 8.3

For measurements of muscle strength to be comparable in repeated strength tests with a dynamometer, the point of application and direction of the force produced by the dynamometer must be recorded. With a Cybex-type apparatus, the dynamometer is aligned with the center of motion of the knee, and the torque is recorded instead of the force. Thus, comparable measurements can be made with the resistance arm placed anywhere along the lower leg.

During isometric muscle testing, the magnitude of the maximum torque produced by muscle forces depends on the joint angle; this angle should also be recorded. If "two-joint" muscles are involved in the strength test, both joint angles should be recorded.

Both the length-tension relationship of the muscle and the moment arm of the muscle force affect the magnitude of the maximum internal torque. In a study of knee extensor muscles, Haffajee et al. (1972) showed that the length-tension relationship of the muscle is at least as significant as the length of the moment arm of the muscle force.

Solutions to Problems 8.4A and B

8.4A. Find the external torque (M) at various knee joint angles when a patient exercises the knee extensor muscles while wearing a 10-kg weight boot.

The solution is given for 30° knee flexion only. Two alternative solutions are presented.

Alternative 1

Graphic Solution

The length of the moment arm for the force produced by the weight boot is designated a, and that for the weight of the leg is designated b. Moment arm lengths a and b can be obtained by constructing two right triangles with hypotenuses of 0.45 m and 0.20 m, respectively, and one acute angle of 30° (compare Figure 8.10.1).

$a = 0.39 \pm 0.01$ m

$b = 0.17 \pm 0.01$ m

$M = 100 \cdot 0.39 + 50 \cdot 0.17$

$M = 48 \pm 2$ Nm

Mathematical Solution

The length of the moment arm (a) for the force produced by the weight boot (T) (compare Figure 8.10.1) is:

$a = 0.45 \cos 30°$

The length of the moment arm (b) for the weight of the leg is:

$b = 0.20 \cos 30°$

$M = 100 \cdot 0.45 \cos 30° + 50 \cdot 0.20 \cos 30°$

$M = 48$ Nm

Alternative 2

Graphic Solution

The force produced by the weight boot (100 N) is resolved into components so that one component (F_t) is perpendicular to the long axis of the tibia, and the other component (P) is parallel to the tibia, passing through the center of motion of the knee (Figure 8.10.2).

$F_t = 87 + 2$ N

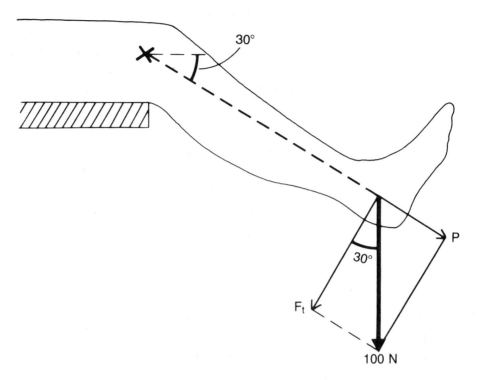

Figure 8.10.2. Resolution of a 100-N force applied to the ankle. The knee is flexed 30°. One component (F_t) is perpendicular to the long axis of the tibia; the other component (P) is parallel to this axis and passes through the center of motion of the knee joint. F_t forms the side of a right triangle with a hypotenuse of 10 kg (100 N) and an acute angle of 30°.

The weight of the lower leg (50 N) is similarly resolved into components, the component perpendicular to the lower leg being designated F_u.

$$F_u = 43 \pm 2 \text{ N}$$

The two components perpendicular to the tibia create a torque on the knee joint, whereas those passing through the center of motion of the knee joint do not. The length of the moment arm for the force component F_t (87 N) is 0.45 m. For the force component F_u (43 N), the moment arm length is 0.20 m.

$$M = 87 \cdot 0.45 + 43 \cdot 0.20$$

$$M = 48 \pm 2 \text{ Nm}$$

Mathematical Solution

The force component (F_t) for the force produced by the weight boot (Figure 8.10.2) is:

$$F_t = 100 \cos 30°$$

The force component (F_u) for the 50-N force produced by the weight of the lower leg is:

$$F_u = 50 \cos 30°$$

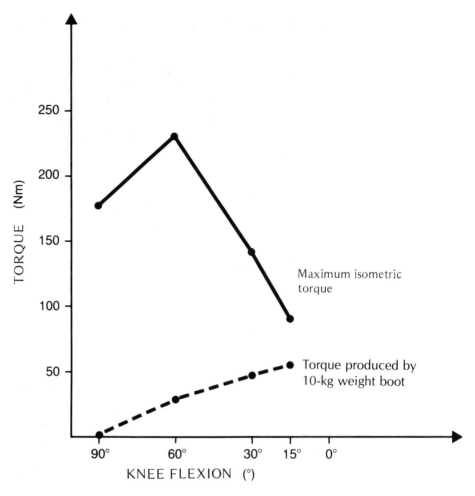

Figure 8.10.3. The torque (Nm) produced by the knee extensor muscles at four angles of knee joint flexion (90°, 60°, 30°, and 15°). The graph shows the maximum isometric torque obtained for a group of young men (Lindahl et al., 1969) as well as the calculated torque produced by a 10-kg (100-N) weight boot. 0° represents the knee fully extended.

M = 100 cos 30° · 0.45 + 50 cos 30° · 0.20

(Compare the moment equation for the mathematical solution given in Alternative 1.)

M = 48 Nm

8.4B. For the solution to this problem, see Figure 8.10.3.

Answers to Discussion Questions in 8.4

1. The torque produced by the weight boot is equal to 12% of the maximum torque at a knee angle of 60°, and 59% of the maximum torque at an angle of 15°. The exercise will be most effective with the knee flexed 15° (Lindh, 1979).

2. The patient may not be able to extend the knee joint fully with a 10-kg weight at the ankle, despite the fact that only 59% of the maximum torque is utilized at a joint angle of 15°. Normally, the maximum torque

over the last 15° of extension is substantially lower than that produced with the knee flexed more than 15°.

In a radiographic study of 52 individuals with healthy knees, Lindahl and Movin (1968) showed that in the supine position, full active extension averages 5° less than full passive knee extension. Therefore, there is a radiographically measurable difference between full normal active extension and full normal passive extension. The torque produced by the weight of the lower leg is only about 10 Nm. Some reasons why full active extension does not equal full passive extension are that during active extension the length-tension relationship of the quadriceps muscle is less favorable, the moment arm of the quadriceps muscle is short, and passive resistance from the knee flexor muscles and posterior joint capsule makes it difficult to extend the knee fully. In an in vitro study, Lieb and Perry (1968) showed that the demand on the quadriceps muscle increases by 60% during the last 15° of knee extension.

Solutions to Problems 8.5A and B

8.5A. Find the external torque on the knee joint when an individual performs knee extensor muscle exercises while seated on the quadriceps table with the knee in full extension (M_1) and in 60° of flexion (M_2). The weight arm is placed parallel to the resistance arm. The 10-kg weight applied to the weight arm produces a vertical force of 100 N. The length of the moment arm is 0.35 m when the knee is in full extension (Figure 8.10.4A), and thus M_1 equals 35 Nm.

Graphic Solution

The length of the moment arm at 60° of knee flexion (Figure 8.10.4B) is obtained by constructing a right triangle with a hypotenuse of 0.35 m and an acute angle of 60° (compare Figure 8.10.1).

$M_2 = 100 \cdot 0.18$

$M_2 = 18 \pm 1$ Nm

Mathematical Solution

The moment arm for the 100-N force produced by the weight attached to the weight arm at 60° of knee flexion is 0.35 cos 60° (Figure 8.10.4B).

$M_2 = 100 \cdot 0.35 \cos 60°$

$M_2 = 18$ Nm

8.5B. Find the force (T) produced by the weight attached to the weight arm when the weight arm is angled upward 60°.

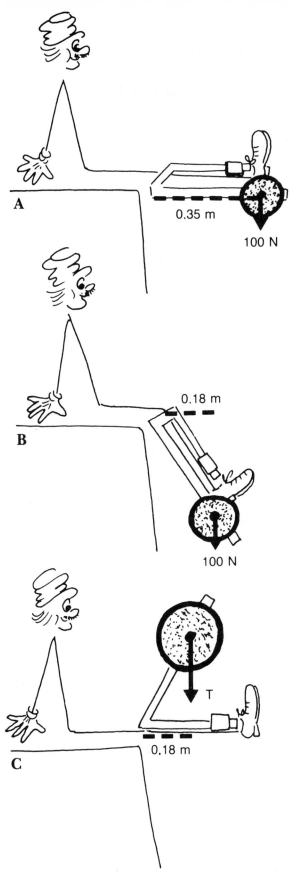

Figure 8.10.4. Knee extensor muscle exercises on the quadriceps table. A. With the weight arm and resistance arm parallel and the knee fully extended, the force produced by the attached weight (100 N) acts with a moment arm of 0.35 m with regard to the center of motion of the knee joint. B. With the knee flexed 60°, the force produced by the attached weight (100 N) acts with a moment arm of 0.18 m. C. With the weight arm angled 60° upward in relation to the resistance arm and the knee fully extended, the force produced by the attached weight (T) acts with a moment arm of 0.18 m.

Graphic Solution

When the weight arm is positioned upward at a 60° angle to the horizontal plane, the moment arm of the force produced by the attached weight (T) is 0.18 ± 0.1 m (Figure 8.10.4C).

$M_1 = 35$ Nm gives the magnitude of force T.

$0.18T = 35$

$T = \dfrac{35}{0.18}$

$T = 194 \pm 10$ N

Mathematical Solution

The length of the moment arm for force T is $0.35 \cos 60°$ (Figure 8.10.4C).

$M_1 = 35$ Nm gives the magnitude of force T.

$T \cdot 0.35 \cos 60° = 35$

$T = 200$ N

Solutions to Problems 8.6A and B

8.6A. Find the patellofemoral force (P) during quadriceps muscle exercises with the knee flexed 5°.

Graphic Solution

Consider the patella as a free body (Figure 8.10.5). The following forces act on the patella: the quadriceps muscle force (Q), the force transmitted through the patellar tendon (T), and the patellofemoral joint reaction force (P).

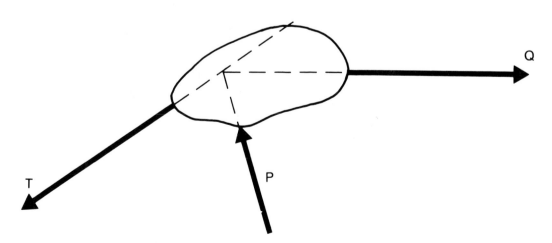

Figure 8.10.5. Free body diagram of the patella. The forces acting on the free body are the quadriceps muscle force (Q), the force transmitted through the patellar tendon (T), and the patellofemoral joint reaction force (P).

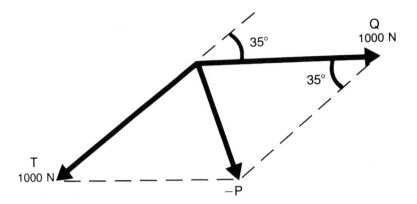

Figure 8.10.6. Parallelogram of forces with two equal components (Q = T = 1000 N) and one outer angle of 35°. The resultant (−P) of the two force components (Q and T) is a diagonal in the parallelogram. The acute angles in the parallelogram are 35°.

Since the patella is assumed to be a frictionless pulley, $Q = T = 1000$ N.

Force P constitutes the counterresultant to forces Q and T and is obtained by means of the parallelogram method (Figure 8.10.6).

$P = 601 \pm 20$ N

(Compare the solution obtained by means of the polygon method in Figure 8.10.15.)

Mathematical Solution

The patellofemoral joint reaction force (P) is obtained from Figure 8.10.6.

The parallelogram of forces is closed (see Figure 8.10.6). The first diagonal (−P) is drawn. The second diagonal is then added. This diagonal forms a right angle with the first diagonal at their intersection point. In the right triangle formed, one acute angle is $\dfrac{35°}{2} = 17.5°$ and the opposite angle is $\dfrac{P}{2}$. In an equilateral parallelogram, the diagonal divides the angle formed by the sides in half. The diagonals then intersect each other at their midpoints at right angles.

$\dfrac{P}{2} = 1000 \sin 17.5°$

$P = 601$ N

8.6B. The solution is similar to that for problem 8.6A.

Figure 8.10.7 illustrates the dependent relationship between the patellofemoral joint reaction force and the knee joint angle.

Answer to Discussion Question in 8.6

The magnitude of the patellofemoral joint reaction force is influenced by the angle between the patellar tendon (T) and the quadriceps tendon

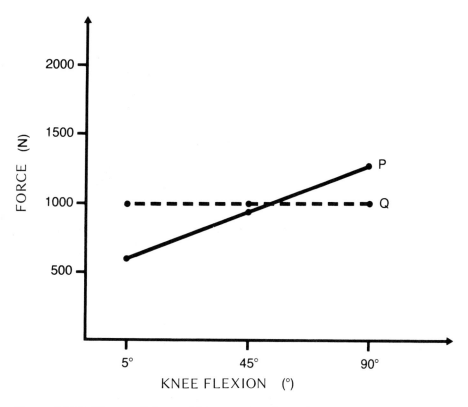

Figure 8.10.7. The patellofemoral joint reaction force (P) at three knee joint angles (5°, 45°, and 90°) with the quadriceps muscle force held constant (Q = 1000 N).

(Q) (Figure 8.6). This angle is influenced in turn by the knee joint angle. Thus, the patellofemoral joint reaction force is dependent on the knee joint angle. As the knee is flexed, the patellofemoral joint reaction force (P) increases even if the quadriceps force (Q) is kept constant (Figure 8.10.7).

The patellofemoral joint reaction force is also affected by the quadriceps muscle force. If the muscle force increases, the patellofemoral joint reaction force also increases. If the quadriceps muscle force is increased by an equal amount at the three joint angles (5°, 45°, and 90°), the increase in magnitude of the joint reaction force will be greater when the knee is flexed than when it is extended.

If the patellofemoral joint reaction force per unit of area is calculated, the size of the contact surface of the joint becomes significant. The size of the contact surface between the patella and the femur changes with the joint angle. When the knee is extended, the lower part of the patella rests against the femur. During knee flexion up to 90°, the contact surface shifts cranially, and the size increases somewhat (Goodfellow et al., 1976). Goodfellow et al. (1976) considered the size of the contact surface to depend more on the angle of knee flexion than on the magnitude of the external torque.

To a certain extent, the increased size of the contact surface compensates for the increased magnitude of the patellofemoral joint reaction force.

Solutions to Problems 8.7A and B

8.7A. Find the external force (F) that must be applied to the ankle of a patient seated with the knee flexed 15° for the quadriceps muscle force (Q) to equal that during a one-leg stance with the knee in 15° of flexion (740 N).

Graphic Solution

The moment arm of force F is equal to the perpendicular distance from force F to the center of motion of the knee joint (compare Figure 8.10.1, in which the knee is flexed 30°). The length of the moment arm is 0.39 ± 0.01 m.

The clockwise moment is positive.

$\Sigma M = 0$ gives the magnitude of the external force F.

$$F \cdot 0.39 - 740 \cdot 0.05 = 0$$

$$F = 96 \pm 3 \text{ N}$$

Mathematical Solution

The moment arm for the external force (F) applied to the ankle is 0.4 cos 15°.

$$F \cdot 0.4 \cos 15° - 740 \cdot 0.05 = 0$$

$$F = 96 \text{ N}$$

8.7B. Find the joint reaction force (R) when the exercise is performed in the standing and sitting positions.

Graphic Solution

The tibiofemoral joint reaction force (R) is first found for the standing position.

The lower leg is considered as a free body with the following forces acting on it: the ground reaction force (700 N) acting vertically upward, the quadriceps muscle force (740 N) acting in a direction of $80° - 20° = 60°$ to the horizontal plane, and the tibiofemoral joint reaction force (R_{stand}) (Figure 8.10.8).

$\Sigma F = 0$ gives the joint reaction force R_{stand}. The magnitude and direction of R_{stand} are obtained by means of the polygon method (Figure 8.10.9).

$$R_{stand} = 1391 \pm 20 \text{ N}$$

The tibiofemoral joint reaction force (R_{sit}) during exercises performed in the sitting position is obtained in a similar manner (Figure 8.10.10). A polygon of forces is constructed. The polygon is composed of the quadriceps muscle force (740 N) acting in a direction of $20° + 15° = 35°$ to the horizontal plane, the vertical external force (96 N), and the tibiofemoral joint reaction force (R_{sit}).

$$R_{sit} = 690 \pm 20 \text{ N}$$

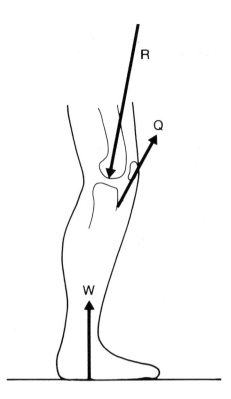

Figure 8.10.8. Quadriceps muscle exercises during a one-leg stance. The following forces act on the lower leg: the body weight (W), the quadriceps muscle force (Q), and the tibiofemoral joint reaction force (R). The weight of the lower leg is disregarded.

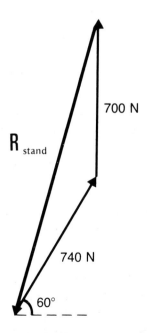

700 N

R_{stand}

740 N

60°

Figure 8.10.9. The polygon of forces reveals the tibiofemoral joint reaction force (R_{stand}) during a one-leg with the knee flexed 15°. The quadriceps muscle force (740 N) forms a 60° angle with the horizontal plane. The ground reaction force (700 N) acts vertically upward.

117

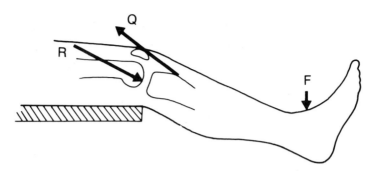

Figure 8.10.10. Quadriceps muscle exercises in a sitting position. The following forces act on the lower leg: an external force (F), the quadriceps muscle force (Q), and the tibiofemoral joint reaction force (R). The weight of the lower leg is disregarded.

Mathematical Solution

The tibiofemoral joint reaction force is first found for the standing position (R_{stand}) (Figures 8.10.8 and 8.10.9) and then for the sitting position (R_{sit}) (Figure 8.10.10).

The tibiofemoral joint reaction force for the standing position (R_{stand}) is obtained from Figure 8.10.9 by means of the cosine theorem. The angle opposite R_{stand} is 150°. An auxiliary right triangle is constructed by drawing a vertical line from the tip of the quadriceps muscle force (740 N) to the horizontal line in the figure. In the triangle, one acute angle is 60°. The other acute angle must then be 30° and is the supplemental angle to the angle sought.

$$R_{stand}^2 = 700^2 + 740^2 - 2 \cdot 700 \cdot 740 \cdot \cos 150°$$

$$R_{stand}^2 = 700^2 + 740^2 + 2 \cdot 700 \cdot 740 \cdot \cos 30°$$

$$R_{stand} = 1391 \text{ N}$$

The tibiofemoral joint reaction force (R_{sit}) for the sitting position is obtained in a similar manner.

In the triangle of forces, the magnitude and direction of the forces composing two sides are known. The quadriceps muscle force of 740 N forms a 35° angle with the horizontal plane, and the external force of 96 N acts vertically. The third side (R_{sit}) is obtained by means of the cosine theorem. The angle opposite R_{sit} is 55°. An auxiliary right triangle is constructed by drawing a vertical line from the tip of the quadriceps muscle force (Q) to a horizontal line extending from the base of force Q (Figure 8.10.10). One acute angle in the auxiliary triangle is formed by the tensile force (35°) produced by the quadriceps muscle. The acute angle sought is then $180° - 35° - 90° = 55°$.

$$R_{sit}^2 = 96^2 + 740^2 - 2 \cdot 96 \cdot 740 \cdot \cos 55°$$

$$R_{sit} = 689 \text{ N}$$

Answer to Discussion Question in 8.7

One advantage of performing quadriceps muscle exercises in a standing position is that the exercises are more physiologic. During these exercises, the quadriceps muscle is being strengthened, and its coordination with other muscles is simultaneously being improved. Better coordination leads to a more favorable load distribution in the joint and probably also results in lower energy consumption.

When the quadriceps muscle force is held constant, the tibiofemoral joint reaction force is smaller during sitting than during standing because the weight of the body above the knee does not act on the knee joint. Therefore, exercising in the sitting position can also be advantageous. In this example, the magnitude of the tibiofemoral joint reaction force during exercises in a sitting position is one half that during exercises in a standing position.

As a rule, very little quadriceps muscle force is required during standing. Thus, if vigorous strengthening exercises are performed in a sitting position, the magnitude of the tibiofemoral joint reaction force equals or exceeds that during moderate exercises in a standing position.

Solutions to Problems 8.8A and B

8.8A. Find the quadriceps muscle force (Q) and the tibiofemoral joint reaction force (R) when an external force of 150 N is applied to the lower leg during quadriceps muscle exercises performed in a sitting position with the knee flexed 30°.

Graphic Solution

Find the quadriceps muscle force (Q).
The counterclockwise moment is positive.

$\Sigma M = 0$ gives the magnitude of the quadriceps muscle force Q.

$0.05Q - 0.4 \cdot 150 = 0$

$0.05Q = 60$

$Q = 1200$ N

Find the tibiofemoral joint reaction force (R).
The following forces act on the lower leg: the quadriceps muscle force (Q), an externally applied force (150 N), and the tibiofemoral joint reaction force (R). The line of application for force R passes through the center of motion of the knee joint.

The quadriceps muscle force (Q) is 1200 N and forms an angle of $20° + 30° = 50°$ with the horizontal plane. The external force is 150 N and forms a 60° angle with the horizontal plane. (The knee is flexed 30°, and the external force is applied perpendicular to the long axis of the tibia.)

$\Sigma F = 0$ gives the tibiofemoral joint reaction force R. The magnitude and direction of force R are found by means of the polygon method (Figure 8.10.11).

$R = 1157 \pm 20$ N

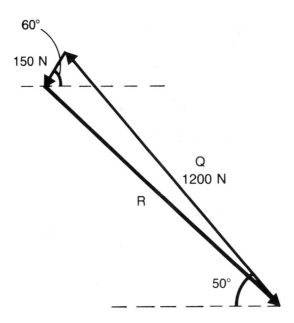

Figure 8.10.11. The tibiofemoral joint reaction force (R) during quadriceps muscle exercises performed in the sitting position with the knee flexed 30° is found by means of a polygon of forces. The quadriceps muscle force (Q = 1200 N) forms an angle of 50° with the horizontal plane. An externally applied force (150 N) forms an angle of 60° with the horizontal plane.

Mathematical Solution

The tibiofemoral joint reaction force (R) is obtained from Figure 8.10.11 by means of the cosine theorem. An auxiliary triangle is constructed by drawing a horizontal line from the tip of the external force of 150 N through force Q. In the triangle thus formed, one angle is 60° and the other is 50°. The angle opposite R is then

$180° - 50° - 60° = 70°$.

$R^2 = 150^2 + 1200^2 - 2 \cdot 150 \cdot 1200 \cdot \cos 70°$

$R = 1157 \text{ N}$

Graphic Solution

Find the shear force (S) on the tibiofemoral joint.

The tibial plateau forms a 60° angle with the horizontal plane. (The knee is flexed 30°, and the tibial plateau is perpendicular to the long axis of the tibia.)

The quadriceps muscle force (Q) is resolved into components, with one component (Q_y) parallel and one component (Q_x) perpendicular to the tibial plateau.

Q_y can be determined graphically, because the magnitude and direction of Q are known. A right triangle is constructed in which the hypotenuse corresponds to 1200 N and the angle opposite Q_y is 20° (Figure 8.10.12).

$Q_y = 410 \pm 30 \text{ N}$

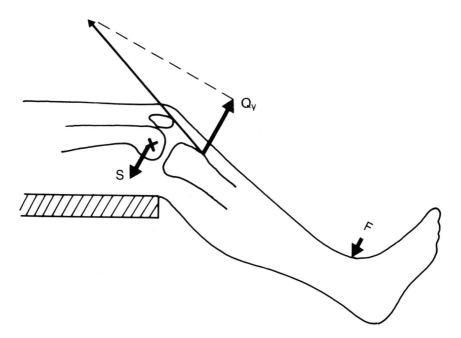

Figure 8.10.12. Quadriceps exercise with the patient in a sitting position and the knee flexed 30°. The quadriceps muscle force component (Q_y) and the external force (F) produced a shear force (S) on the joint.

The forces directed to the left are designated as positive.

$\Sigma F = 0$ gives the magnitude of the shear force S.

$S + 150 - Q_y = 0$

$S + 150 - 410 = 0$

$S = 260 \pm 30$ N

The shear force on the joint is thus directed dorsally (Figure 8.10.12), indicating that a joint reaction force of 260 N prevents the lower leg from being displaced ventrally.

Mathematical Solution

The force component (Q_y) of the quadriceps muscle force that parallels the tibial plateau is obtained from Figure 8.10.12.

$Q_y = 1200 \sin 20°$

$Q_y = 410$ N

$S + 150 - 410 = 0$

$S = 260$ N

8.8B. Find the quadriceps muscle force (Q) when an external force of 150 N is applied to the lower leg during quadriceps muscle exercises with the knee flexed 90°.

The solution is similar to that for problem 8.8A.
Find the tibiofemoral joint reaction force (R).
Again, the solution is similar to that for problem 8.8A.

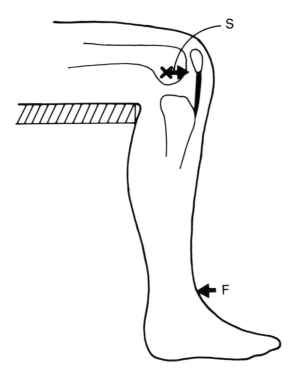

Figure 8.10.13. Quadriceps muscle exercise performed in the sitting position with the knee flexed 90°. The magnitude of the shear force (S) is dependent on the magnitude of the externally applied force (F).

Find the shear force (S) in the tibiofemoral joint.
The forces directed to the left are designated as positive.

$\Sigma F = 0$ gives the magnitude of the shear force S.

$S + 150 = 0$

(Force Q acts vertically and has no horizontal component parallel to the tibial plateau.)

$S = -150 \text{ N}$

The shear force on the joint is thus directed ventrally (Figure 8.10.13). To counteract this shear force, the soft tissues of the knee joint must absorb a force of 150 N to prevent the lower leg from being displaced dorsally.

Answers to Discussion Questions in 8.8

1. If the quadriceps muscle force is held constant, the magnitude of the tibiofemoral joint reaction force is not appreciably affected by the degree of knee flexion. (This situation differs from that in the patellofemoral joint; see example 8.6.) The tibiofemoral joint reaction force is affected primarily by the quadriceps muscle force; thus, a strong muscle force produces a large joint reaction force. The direction of the tibiofemoral joint reaction force is altered by a change in joint angle, however, and thus the distribution of the load in the joint is affected.

2. In this example, performing quadriceps muscle exercises with the knee flexed 30° subjects the anterior cruciate ligament to a load of 260 N, and performing the exercises with the knee flexed 90° subjects the posterior cruciate ligament to a load of 150 N.

During the clinical drawer test, an external force of up to approximately 100 N is applied to the knee joint (Markolf et al., 1976; Markolf et al., 1978). Both during stair climbing and during walking on level ground, the anterior cruciate ligament is subjected to a tensile force of 100 N to 200 N (Morrison, 1969). During moderate to vigorous physical activity, the knee joint is subjected to shear forces of 300 N to 500 N (Butler et al., 1980). By comparison, the shear force of 260 N produced by the quadriceps muscle during strengthening exercises performed in the sitting position with the knee flexed 30° can be considered significant.

The anterior cruciate ligament is under tension from full extension of the knee to 40° of flexion (Detenbeck, 1974). Contraction of the quadriceps muscle with the knee in this range of motion increases the tensile load on this ligament because the muscle force tends to augment the tensile force on the ligament. Therefore, it is theoretically correct to restrict vigorous exercises in this range of knee motion after anterior cruciate ligament injury, as a year may be needed before this ligament regains its full strength (Noyes, 1977).

With the knee flexed 50–90°, however, the quadriceps muscle force does not appreciably increase the tensile load on the anterior cruciate ligament. Hence, it is theoretically defensible to exert a larger external torque on the lower leg when the patient performs quadriceps muscle exercises in the sitting position. However, few studies have examined the stability of the knee joint after the use of various treatment modalities in the rehabilitation of a patient with an anterior cruciate ligament injury.

It should be pointed out that in this example the counteracting torque produced by the cruciate ligament is not considered in the calculation of the quadriceps muscle force (Q). Under tension, the anterior cruciate ligament exerts a flexing torque on the knee joint and the posterior cruciate ligament exerts an extending torque. In this example, the magnitudes of these torques are negligible.

Solutions to Problems 8.9A and B

8.9A. Find the tibiofemoral joint reaction force (R) during stair climbing and descending. The solution is given for climbing only.

Graphic Solution

Three forces act on the lower leg: the ground reaction force (700 N), the force transmitted through the patellar tendon (T), and the tibiofemoral joint reaction force (R).

The counterclockwise moment is positive.

$\Sigma M = 0$ gives the magnitude of force T.

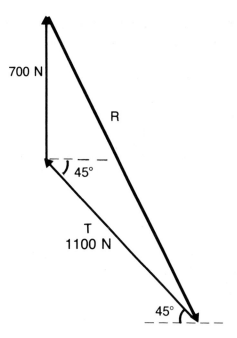

Figure 8.10.14. The polygon of forces is used to find the tibiofemoral joint reaction force (R) during stair climbing. The patellar tendon force (T = 1100 N) forms a 45° angle with the horizontal plane. The ground reaction force (700 N) acts vertically upward.

$55 - 0.05\ T = 0$

$T = 1100\ N$

The direction of T is $60 - 15 = 45°$ to the horizontal plane. (The long axis of the tibia slopes 60°, and the patellar tendon forms a 15° angle with this axis.)

$\Sigma F = 0$ gives the tibiofemoral joint reaction force R. The magnitude and direction of force R are found by means of the polygon method (Figure 8.10.14).

$R = 1670 \pm 20\ N$

The line of application of force R passes through the center of motion of the knee joint.

Mathematical Solution

The tibiofemoral joint reaction force (R) during stair climbing is found by means of the cosine theorem (see Figure 8.10.14). The angle opposite R is $90° + 45° = 135°$.

$R^2 = 1100^2 + 700^2 - 2 \cdot 1100 \cdot 700 \cdot \cos 135°$

$R^2 = 1100^2 + 700^2 + 2 \cdot 1100 \cdot 700 \cdot \cos (180° - 135°)$

$R = 1670\ N$

8.9B. Find the patellofemoral joint reaction force (P) during stair climbing and descending. The solution is given for stair climbing only.

Graphic Solution

Three forces act on the patella: the force transmitted through the patellar tendon (T), the quadriceps muscle force (Q), and the patellofemoral joint reaction force (P).

The assumption that the patella is a frictionless pulley means that Q=T=1100 N, and force T acts at a 45° angle to the horizontal plane (from problem 8.9A).

To find the direction of force Q, an auxiliary horizontal line is drawn through the knee joint depicted in Figure 8.9A. The 60° angle designated in the figure, between the long axis of the tibia and the horizontal plane, is the alternative angle to the angle at the level of the knee joint. The remaining angle between the long axis of the femur and the horizontal plane is thus $130° - 60° = 70°$.

$\Sigma F = 0$ gives the patellofemoral joint reaction force P. The magnitude of force P is obtained by means of the polygon method (Figure 8.10.15).

$$P = 1182 \pm 30 \text{ N}$$

(For an alternate solution, see the solution to problem 8.6A.)

Mathematical Solution

The patellofemoral joint reaction force (P) during stair climbing is found by means of the cosine theorem (see Figure 8.10.15). The angle opposite force P is 65°.

$$P^2 = 2 \cdot 1100^2 - 2 \cdot 2 \cdot 1100 \cdot \cos 65°$$

$$P = 1182 \text{ N}$$

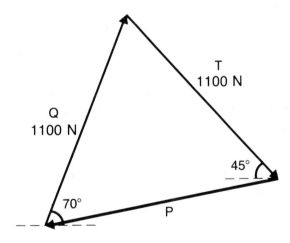

Figure 8.10.15. The polygon of forces is used to find the patellofemoral joint reaction force (P) during stair climbing. The force through the patellar tendon (T = 1100 N) forms a 45° angle with the horizontal plane, and the quadriceps muscle force (Q = 1100 N) forms a 70° angle with this plane.

8.11. Commentary

In three examples involving the knee (8.1, 8.3, and 8.4), the values for the maximum isometric torque produced by the knee extensor muscles are taken from several studies of normal young men. Figure 8.11.1 com-

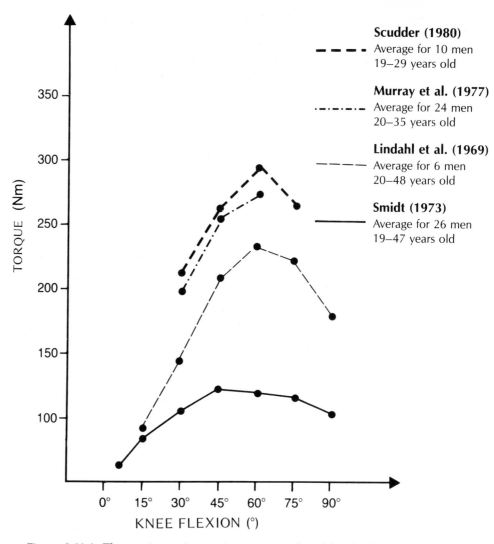

Scudder (1980)
‑ ‑ ‑ Average for 10 men
19–29 years old

Murray et al. (1977)
·‑··‑··‑ Average for 24 men
20–35 years old

Lindahl et al. (1969)
‑ ‑ ‑ Average for 6 men
20–48 years old

Smidt (1973)
——— Average for 26 men
19–47 years old

Figure 8.11.1. The maximum isometric torque produced by the knee extensor muscles as a function of the knee joint angle. The curves represent data from four studies of young healthy men.

pares the values for the maximum isometric torque produced by the knee extensor muscles from four studies (Lindahl et al., 1969; Murray et al., 1977; Scudder, 1980; Smidt, 1973). The shapes of the curves for the maximum isometric torque are similar in these studies, with peak magnitudes occurring at 50–60° of knee flexion. The values for the torques differ widely among the studies, however. Smidt's study (1973) shows very low values throughout the range of knee motion for men from 19 to 47 years of age. These values are considerably lower than values for men aged 45 to 65 years (Murray et al., 1977), which are not shown in the figure. An explanation for this difference may be that the patients in Smidt's study were lying on their sides with their hip joints fully extended, whereas in the other studies the patients were in the sitting position. The studies conducted by Scudder (1980) and Murray et al. (1977) made use of the Cybex II test apparatus; the results from these two studies are in close agreement. If one or more heads of the quadriceps muscle have been surgically removed, the reduction in muscle strength is greatest in the range of knee motion where the largest moment is normally exerted on the joint. The reduction in muscle strength is less pronounced when the knee is extended further or flexed further (Markhede, 1980).

The Torque Produced by the Knee Extensor Muscles during Physiologic Exercises

During walking, the average maximum flexing torque on the knee joint varies between 2% and 4% of the body weight multiplied by the individual's height. The value for the maximum flexing torque is partly dependent on the walking speed (Andriacchi et al., 1981). (For a comparison of the torque on the hip joint, see Chapter 7, Section 7.10, Commentary.) The portion of the gait cycle during which the torque on the knee joint reaches its maximum value is relatively short, occurring in the first half of the stance phase when the knee is flexed and then extended again.

During gait, the interindividual pattern of motion in the knee joint varies more than does that in the hip joint. During slow walking, some subjects keep the knee almost fully extended throughout the entire stance phase, and thus they do not use the knee extensor muscles to any appreciable extent. During fast walking, the interindividual pattern of motion in the knee joint tends to become more uniform.

When a person kicks a ball, the force produced by the quadriceps muscle is about 3000 N (Frankel and Burstein, 1970). If the moment arm of the quadriceps muscle force is 0.05 m, the torque on the knee joint is about 150 Nm. Wahrenberg et al. (1978) found that the torque on the knee joint during kicking of a ball can reach a value of 260 Nm.

The external torque produced during weight lifting was studied in the hip, knee, and ankle. One subject ruptured a patellar tendon when he lifted a barbell weighing 175 kg. At the instant that the tendon ruptured, the knee was flexed 90°, and the torque on the knee joint was 550 Nm. The tendon ruptured about 0.4 second after the lift was begun (Zernicke et al., 1977).

During physiologic exercise, the flexing torque on the knee joint varies between 20 Nm during walking on level ground (for a person with a weight of 60 kg and a height of 1.65 m) and the extreme case of 550 Nm when an individual lifts weights vigorously.

Loading of the Knee Joint

During gait, the largest compressive force on the tibiofemoral joint is produced by the gastrocnemius muscle. During push off, this force may reach a value of four times body weight (Morrison, 1970). Calculation of this value disregards the forces produced by the antagonistic muscles. A different method of calculation, which takes into account several muscle groups, shows that the knee joint reaction force can reach seven times body weight during level walking (Seireg and Arvikar, 1975). During stair climbing, the compressive force on the tibiofemoral joint is calculated to be about four times body weight (Morrison, 1969).

The patellofemoral joint reaction force during walking on level ground is 0.5 times body weight, and during stair climbing, three times body weight (Reilly and Martens, 1972). The value of three times body weight corresponds to the value obtained in example 8.9 for the patellofemoral joint reaction force during stair climbing.

Assumptions Made in Examples 8.1 through 8.8

1. In the examples, the center of motion of the knee joint is placed within the area wherein the anatomic center of motion shifts throughout the range of knee motion. The shifting of the anatomic instantaneous center of motion in the sagittal plane is determined by means of the instant center technique described by Reuleaux (Frankel and Nordin, 1980).

In the literature, the placement of the center of motion of the knee joint is found to vary for calculations of forces acting on the knee joint in the sagittal plane. The center of motion is commonly placed in the joint space between the femur and tibia, which is not within the region of shift of the anatomic instantaneous center of motion of the knee joint.

In a three-dimensional analysis of the forces acting on the knee joint during walking, Morrison (1968) placed the center of motion at the center of the tibial plateau. Nissan (1980) analyzed some of the assumptions made by Morrison and showed that during walking, the placement of the center of motion in the ventrodorsal direction greatly affected the calculated force values. Shifting the center of motion 1 cm ventrally or 1 cm dorsally was found to alter the values significantly. However, a craniocaudal shift of 1 cm did not affect the values.

2. The moment arm of the quadriceps muscle force is assumed to be 0.05 m. In the range of motion of 0–90° of knee flexion, the length of the moment arm of the quadriceps muscle force varies between 3.8 and 4.90 cm (Smidt, 1973). Uncertainties in the measurements are so great, however, that the moment arm length can be expressed in whole centimeters, i.e., 4 to 5 cm.

3. The patella is assumed to be a frictionless pulley (Matthews et al., 1977; Reilly and Martens, 1972). A lack of friction is assumed both between the quadriceps tendon and the patella and between the patella and the femur. Denham and Bishop (1978) claimed that friction can be disregarded between the patella and the femur, but not between the patella and the quadriceps tendon. Therefore, the patellar tendon force cannot be considered to be as great as the quadriceps muscle force. In examples 8.6 and 8.9, friction between the tendon and the patella is disregarded in order to simplify the calculations.

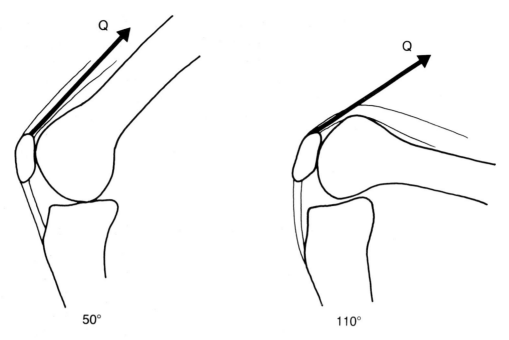

50°   110°

Figure 8.11.2. The position of the quadriceps tendon relative to the long axis of the femur. At 50° of knee flexion, the force transmitted through the quadriceps tendon (Q) can be considered to parallel the long axis of the femur. At 110° of knee flexion, the quadriceps tendon curves around the femoral condyles, and force Q no longer parallels the long axis of the femur (From Bandi, 1977).

4. The muscle force transmitted through the quadriceps tendon is assumed to parallel the long axis of the femur (Reilly and Martens, 1972). This assumption is reasonable within the range of motion of 0–50° of knee flexion. When the knee is flexed beyond 50°, the patella slides between the femoral condyles, causing the quadriceps tendon to curve around the femoral condyles (Bandi, 1977) (Figure 8.11.2). Thus, the force transmitted through the quadriceps tendon can no longer be considered to parallel the femur. During stair descension in example 8.9B, the knee is flexed 65°; hence, use of this assumption probably means that the value for the patellofemoral joint reaction force is rather high.

5. The anterior cruciate ligament is assumed to counteract ventrally directed shear forces, and the posterior cruciate ligament is assumed to counteract dorsally directed shear forces. The anterior and posterior drawer test was performed in an in vitro study at 30° and 90° of knee flexion (Butler et al., 1980). With the ligaments intact, an average of 440 N was required at 90° of knee flexion, and 333 N at 30° of knee flexion, to displace the tibia 5 mm relative to the femur. When the anterior drawer test was performed at 30° of knee flexion, the anterior cruciate ligament absorbed 87% of the load. When the posterior drawer test was performed at 90° of knee flexion, the posterior cruciate ligament absorbed 94% of the force. This study showed that the cruciate ligament counteracts almost all of the shear forces in the knee joint in the absence of stabilizing activity of the hamstring muscles.

9

The Lumbar Spine

The lumbar spine sustains the greatest loads of any part of the vertebral column. This area of the spine consists of five motion segments, L1 to L5. One motion segment is composed of two vertebrae and their intervening soft tissues. In the motion segment, the disc and vertebral bodies absorb the largest portion of the loads on the lumbar spine.

All of the examples in this chapter analyze the lumbar spine in the sagittal plane, because the lumbar spine has the greatest mobility and sustains the largest torque in this plane. In a normal lumbar spine, the transverse axis of motion passes through the disc in each motion segment (Rolander, 1966).

9.1. The Erector Spinae Muscle Force during Lifting

Problem

Torque

A 10-kg box is to be lifted onto a high shelf. The erector spinae muscles are assumed to counteract the entire forward-bending moment. How great a force (E) must the erector spinae muscles develop if the moment arm of the force is 0.05 m? The upper body is assumed to weigh 40 kg, and its center of gravity in this case is 0.02 m dorsal to the transverse axis of motion in the L5 disc. The center of gravity of the 10-kg box is 0.3 m ventral to the center of motion in the L5 disc (Figure 9.1).

9.2. Loads on the L5 Disc during Bedmaking

Moment Equilibrium; Force Equilibrium

In this example, the loads on the lumbar spine at the level of L5 during bedmaking are calculated. The example allows a comparison of the loads on the L5 disc during bedmaking with the upper body upright and with the upper body bent forward.

Problems

9.2A. How large is the forward-bending moment (M) on the L5 disc when the individual stands upright (Figure 9.2A) and bends forward (Figure 9.2B) while making a bed? The necessary data for solving the problem can be obtained from the figures.

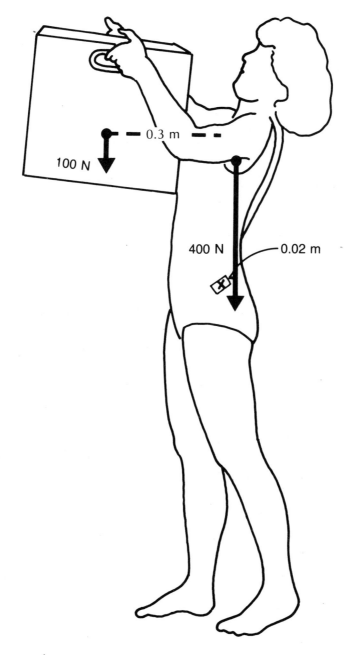

Figure 9.1. Lifting a box onto a high shelf. The weight of the box is 100 N, and its center of gravity is 0.3 m from the center of motion in the L5 disc. The weight of the upper body, including the arms, is 400 N, and its center of gravity is 0.02 m from the center of motion in the L5 disc.

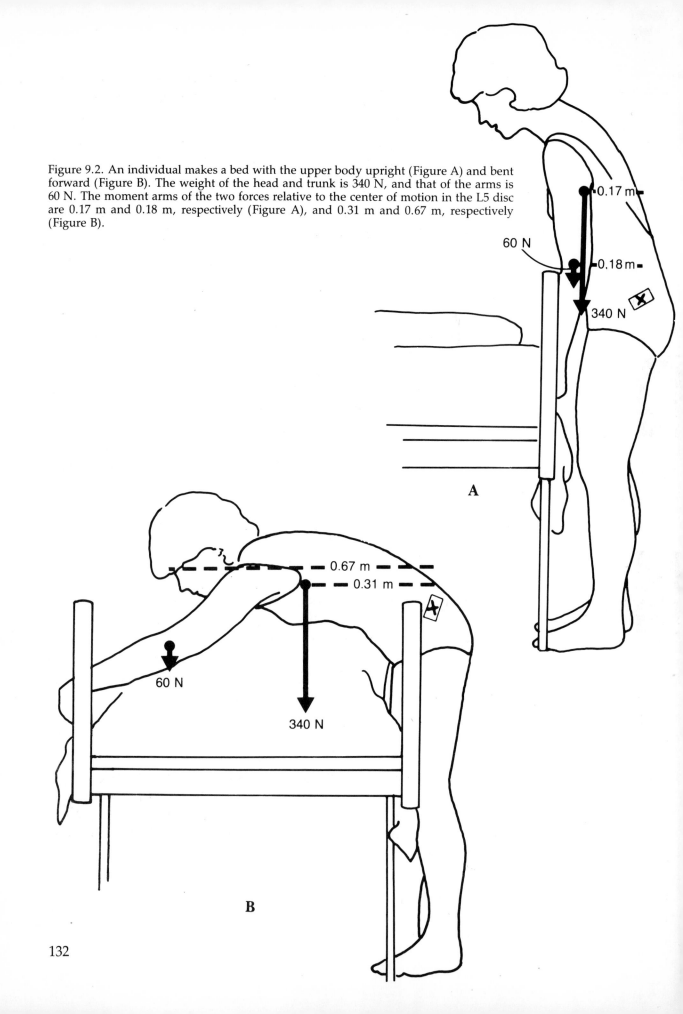

Figure 9.2. An individual makes a bed with the upper body upright (Figure A) and bent forward (Figure B). The weight of the head and trunk is 340 N, and that of the arms is 60 N. The moment arms of the two forces relative to the center of motion in the L5 disc are 0.17 m and 0.18 m, respectively (Figure A), and 0.31 m and 0.67 m, respectively (Figure B).

9.2B. How large a force (E) must the erector spinae muscles exert in each case to counteract the forward-bending moment? The moment arm for the erector spinae muscle force is 0.05 m.

9.2C. How large will the compressive force (C) and the shear force (S) on the L5 disc be in the two cases? The angle of inclination of the L5 disc to the horizontal plane is 30° with the upper body upright (Figure 9.2A) and 70° with the upper body bent forward (Figure 9.2B). The line of application of the erector spinae muscle force is perpendicular to the disc inclination.

Discussion Questions

1. How much larger is the erector spinae muscle force when the individual bends forward than when she stands upright while making the bed? How much larger is the compressive force on the L5 disc?

2. Develop the problem further and discuss the distribution of loads on the disc with the individual in a forward-bending position. How are the loads on the disc distributed if the soft tissues dorsal to the disc fully balance the forward-bending moment? How are the loads distributed if they do not balance the entire moment? Assume in such a case that the vertebral bodies and their intervertebral discs absorb part of the forward-bending moment.

9.3. Loads on the L3 Disc during Static Lifting

Moment Equilibrium; Force Equilibrium

The physical definition of lifting work is introduced in this example.

A biomechanical analysis was carried out in a factory to determine whether a change in the height of a conveyor belt affected the loads on the lumbar spines of workers lifting objects from the conveyor belt onto a cart. When the workers lifted objects repeatedly throughout the day, the lifting frequency was included in the analysis.

Problems

A person working at a 30-cm-high conveyor belt lifted a 9-kg object from the conveyor belt onto a cart 70 cm from the floor about 400 times a day (Figure 9.3A). When the work area was rebuilt, the height of the conveyor belt was changed to 70 cm (to equal the height of the cart) (Figure 9.3B).

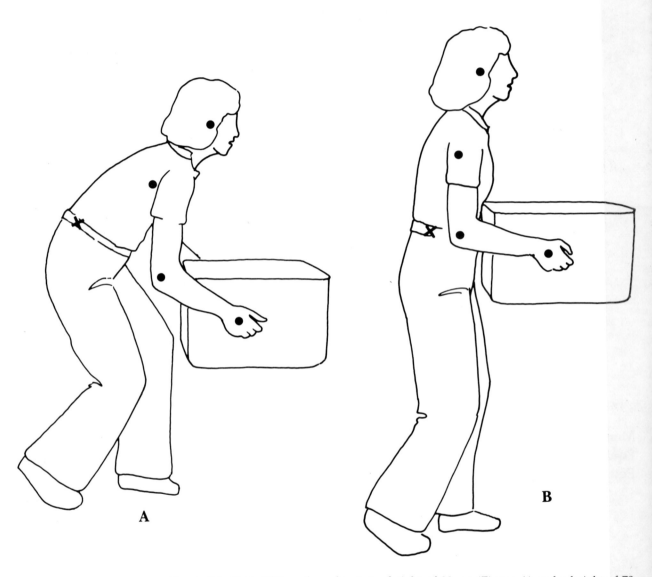

Figure 9.3. Static lifting of an object at a height of 30 cm (Figure A) and a height of 70 cm (Figure B). The centers of gravity of pertinent body segments are marked with solid dots. The center of motion of the L3–L4 disc is marked in each figure with an X.

Use the following data to solve the problems:

Body weight (W)	58 kg
Weight of the head	5.0% of W
Weight of the arms	8.9% of W
Weight of the trunk	36.1% of W
Weight of the object	9 kg

Force	Moment Arm (cm) Fig. 9.3A	Fig. 9.3B
Force produced by weight of head	42.2	17.4
Force produced by weight of trunk	21.3	10.4
Force produced by weight of arms	35.3	22.0
Force produced by weight of object	52.2	38.3
Erector spinae muscle force at L3 level	4.0	4.0
Line of application of erector spinae muscle force (d) relative to the vertical plane at L3 level	44°	14°

The line of application of the erector spinae muscle force is perpendicular to the L3 disc inclination in both lifting situations. In these problems, the center of motion is located in the L3 disc.

9.3A. How much lifting work (L) is carried out in one day in the two cases?

9.3B. Calculate the reaction force (R) on the L3 disc. Assume that the erector spinae muscles are the only active muscle group and that they absorb the entire forward-bending moment. The activity of the abdominal muscles is disregarded.

9.3C. Calculate the compressive force (C) and the shear force (S) on the L3 disc in the two lifting situations.

Discussion Questions

1. In which case is the most lifting work done, moving the object from the 30-cm-high conveyor belt to the cart or from the 70-cm-high belt to the cart?

2. What are the three most important factors affecting the magnitude of the loads on the lumbar spine when an object is lifted?

9.4. Answers, Solutions, and Discussion

Answers to the Problems

9.1.   E = 440 N

9.2A. M (upright) = 68.6 Nm
M (bending forward) = 145.6 Nm

9.2B. E (upright) = 1372 N
E (bending forward) = 2912 N

9.2C.   *Upright*   *Bending Forward*

C = 1718 ± 10 N   C = 3049 ± 10 N
S = 200 ± 10 N   S = 376 ± 10 N

9.3A. L (30-cm-high belt) = 14,400 Nm/day
L (70-cm-high belt) = 0 Nm/day

9.3B. R (30-cm-high belt) = 3333 ± 40 N
R (70-cm-high belt) = 2185 ± 40 N

9.3C.   *30-cm-High Belt*   *70-cm-High Belt*

C = 3322 ± 10 N   C = 2183 ± 10 N
S = 263 ± 10 N   S = 92 ± 10 N

The ranges in some of these answers allow for a reasonable margin of error in the graphic solution of the problems. These ranges were determined on the basis that 2 cm represents 100 N in problems 9.2 and 9.3C, that 1 cm represents 200 N in problem 9.3B, and that the maximum allowable error in measurement is 1 mm of length and 1° of angle.

Solution to Problem 9.1

Find the erector spinae muscle force (E) when a 10-kg box is lifted onto a high shelf. In this problem, the counterclockwise moment is positive.

$\Sigma M = 0$ gives the magnitude of the erector spinae muscle force E.

$100 \cdot 0.3 - 400 \cdot 0.02 - E \cdot 0.05 = 0$

$30 - 8 = 0.05E$

$E = 440$ N

Solutions to Problems 9.2A–C

The calculations below relate to bedmaking with the upper body in the upright position (Figure 9.2A).

9.2A. Find the forward-bending moment (M) on the L5 disc (Figure 9.2A). In this problem, the counterclockwise moments are positive.

$M = 340 \cdot 0.17 + 60 \cdot 0.18$

$M = 68.6$ Nm

9.2B. Find the erector spinae muscle force (E). In this problem, the clockwise moment is positive.

$\Sigma M = 0$ gives the magnitude of the erector spinae muscle force E.

$E \cdot 0.05 - 68.6 = 0$

$$E = \frac{68.6}{0.05}$$

$$E = 1372 \text{ N}$$

9.2C. Find the compressive force (C) and the shear force (S) on the L5 disc.

The line of application of the erector spinae muscle force (E) (1372 N) is perpendicular to the disc inclination, which means that the force acts compressively. The weight of the head and trunk and the weight of the arms act vertically, and the disc is oriented at a 30° angle to the horizontal plane. The components of the weight of the head and trunk and of the arms (340 N + 60 N) acting perpendicularly (C_w) or tangentially (S) to the disc inclination are obtained by resolving the forces (Figure 9.4.1). The resolution of forces results in a right triangle in which one acute angle is 30° and the hypotenuse corresponds to 400 N. The compressive force (C_w) forms one side of the triangle, and the shear force (S) constitutes the other.

$$S \quad = 200 \pm 10 \text{ N}$$

$$C_w = 346 \pm 10 \text{ N}$$

$$C \quad = E + C_w$$

$$C \quad = 1372 + 346$$

$$C \quad = 1718 \pm 10 \text{ N}$$

The shear force (S) is obtained from Figure 9.4.1.

$$S = 400 \sin 30°$$

$$S = 200 \text{ N}$$

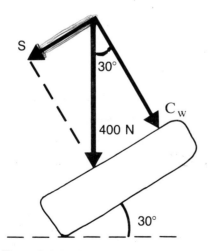

Figure 9.4.1. The L5 disc is oriented at a 30° angle to the horizontal plane. The components of the arm weight and the head and trunk weight (400 N) acting perpendicularly (C_w) and tangentially (S) to the L5 disc inclination are obtained from the parallelogram of forces in which the diagonal corresponds to 400 N and one acute angle is 30°. (Scale: 100 N = 1 cm.)

The force component (C_w) produced by the weight of the upper body, acting perpendicularly to the disc, is obtained from Figure 9.4.1.

$C_w = 400 \cos 30°$

$C_w = 346$ N

The compressive force (C) acting on the L5 disc is equal to the sum of force E and component C_w (E is obtained from problem 9.2B).

$C = 1372 + 346$

$C = 1718$ N

Answers to Discussion Questions in 9.2

1. During bedmaking, the erector spinae muscle force is 1540 N larger when the individual bends forward than when she maintains an upright position, and the compressive force is 1331 N larger.

2. If the forward-bending moment is fully balanced by the soft tissues dorsal to the disc, only compressive stress is sustained by the disc (Figure 9.4.2A). If the disc absorbs part of the forward-bending moment, however, tensile stress occurs on the dorsal side of the disc, and compressive stress occurs on the ventral side (Figure 9.4.2B). The distribution of loads on the disc therefore depends on which structures absorb the forward-bending moment and may not directly depend on the amount of flexion of the spine. The distribution of loads on the disc cannot be determined in vivo with the measurement methods presently available.

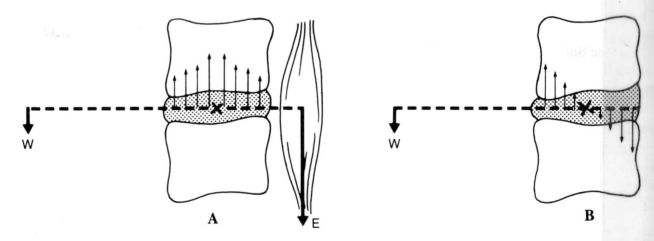

Figure 9.4.2. Two vertebral bodies and their intervertebral disc. The center of motion is marked with an X. The weight of the upper body (W) produces a forward-bending moment on the motion segment. A. If the forward-bending moment produced by the weight of the upper body (W) is primarily absorbed by the erector spinae muscle force (E) on the dorsal side of the disc, then compressive stress occurs on the disc. B. If the forward-bending moment produced by the weight of the upper body (W) is absorbed mainly by the disc and the vertebral bodies, then tensile stress develops dorsally, and compressive stress develops ventrally with respect to the center of motion in the disc.

Solutions to Problems 9.3A–C

9.3A. Find the lifting work (L) carried out in one day by a worker repeatedly moving an object from a 30-cm-high conveyor belt and from a 70-cm-high conveyor belt to a 70-cm-high cart.

Lifting work is defined by the formula $L = mg \cdot h \cdot r$, where mg is the weight of the object, h is the vertical path over which the object is carried, and r is the number of lifts.

If an object is moved without friction from a lower to a higher level, the work is considered to be lifting work (regardless of the path over which the object is carried). Lifting work is distinguished from carrying work in that only vertical motion is considered. By contrast, in this context carrying work involves horizontal motion. Lifting work and carrying work often occur simultaneously.

With the 30-cm-high conveyor belt, the difference in height with respect to the cart is 40 cm. Thus, the lifting work (L) carried out in one day is:

$$L = 90 \cdot 0.40 \cdot 400$$

$$L = 14,440 \text{ Nm/day}$$

With the rebuilt, 70-cm-high conveyor belt, the difference in height with respect to the cart is 0 cm. Thus, the lifting work (L) carried out in one day is:

$$L = 90 \cdot 0 \cdot 400$$

$$L = 0 \text{ Nm/day}$$

No lifting work is performed.

9.3B. Find the reaction force on the L3 disc for the two lifting situations.

Graphic Solution

In this problem, the clockwise moments are positive.

The sum of the moments produced by the weight of pertinent body segments and the object lifted gives the total forward-bending moment for the two situations.

	30-cm-High Belt	70-cm-High Belt
	Figure 9.3A	Figure 9.3B
Moment produced by weight of head	$29 \cdot 0.422 = 12.2$ Nm	$29 \cdot 0.174 = 5.0$ Nm
Moment produced by weight of arms	$52 \cdot 0.353 = 18.3$ Nm	$52 \cdot 0.220 = 11.4$ Nm
Moment produced by weight of trunk	$209 \cdot 0.213 = 44.5$ Nm	$209 \cdot 0.104 = 21.7$ Nm
Moment produced by weight of object lifted	$90 \cdot 0.522 = 47.0$ Nm	$90 \cdot 0.383 = 34.5$ Nm
Total forward-bending moment	122.0 Nm	72.6 Nm

$\Sigma M = 0$ gives the magnitude of the erector spinae muscle force E.

	30-cm-High Belt Figure 9.3A	70-cm-High Belt Figure 9.3B
	$122.0 - E \cdot 0.04 = 0$ $E = 3050$ N	$72.6 - E \cdot 0.04 = 0$ $E = 1815$ N

$\Sigma F = 0$ gives the magnitude of the reaction force R on the L3 disc.

The combined weight of the upper body and lifted object (379 N) acts vertically. The line of application of the erector spinae muscle force forms a 44° angle with the vertical plane with the individual in the more forward-bent position (Figure 9.4.3A). With the individual in the more upright position, the line of application forms a 14° angle with this plane (Figure 9.4.3B). The reaction force (R) is obtained graphically by means of the polygon method.

	30-cm-High Belt Figure 9.4.3A	70-cm-High Belt Figure 9.4.3B
	$R = 3333 \pm 40$ N	$R = 2185 \pm 40$ N

Mathematical Solution

The reaction force (R) on the L3 disc when the conveyor belt is 30 cm high is obtained from Figure 9.4.3A by means of the cosine theorem.

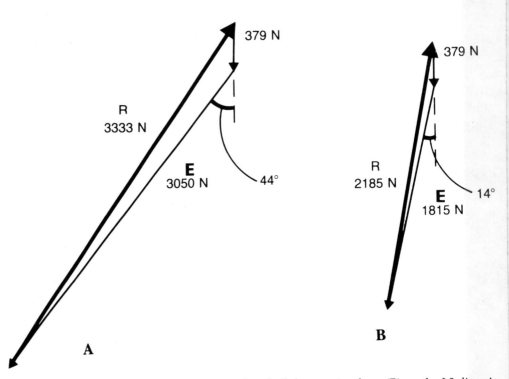

Figure 9.4.3. A polygon of forces is used to find the reaction force (R) on the L3 disc. A. During a lift from a 30-cm-high conveyor belt, the line of application of the erector spinae muscle force (E) forms a 44° angle with the vertical plane at the L3 level. The magnitude of this force is 3050 N. The weight of the upper body (379 N) acts vertically. The reaction force (R) on the L3 disc is 3333 N. B. During a lift from a 70-cm-high conveyor belt, the line of application of force E forms a 14° angle with the vertical plane and acts with a magnitude of 1815 N. The reaction force (R) is 2185 N.

$R^2 = 379^2 + 3050^2 - 2 \cdot 379 \cdot 3050 \cdot \cos(180° - 44°)$

$R^2 = 379^2 + 3050^2 + 2 \cdot 379 \cdot 3050 \cdot \cos 44°$

$R = 3333$ N

The reaction force (R) on the L3 disc when the conveyor belt is 70 cm high is obtained in the same manner from Figure 9.4.3B.

$R^2 = 379^2 + 1815^2 - 2 \cdot 379 \cdot 1815 \cdot \cos(180° - 14°)$

$R^2 = 379^2 + 1815^2 + 2 \cdot 379 \cdot 1815 \cdot \cos 14°$

$R = 2185$ N

9.3.C. Find the compressive force (C) and the shear force (S) on the L3 disc during static lifting of an object from a 30-cm-high conveyor belt and a 70-cm-high conveyor belt.

Graphic Solution

The lines of application of the erector spinae muscle force (E) and the combined weight of the upper body and lifted object (379 N), as well as the angle of inclination of the L3 disc to the transverse plane for both lifting situations, are obtained from Figures 9.4.3A and B.

The compressive force (C) and the shear force (S) for the two lifting situations are obtained by resolution of forces. For the lift from a 30-cm height, a right triangle is constructed with an acute angle of 44° and a hypotenuse corresponding to 379 N. For the lift from a 70-cm height, the right triangle has one acute angle of 14° and a hypotenuse corresponding to 379 N. (Compare Figure 9.4.1.)

Mathematical Solution

The shear force (S) and the compressive force (C) on the L3 disc are obtained through a method similar to that used in problem 9.2C. (See Figure 9.4.1.) The angle of 30° in Figure 9.4.1 changes in this example to 44° when the conveyor belt is 30 cm high, and to 14° when the belt is 70 cm high. The upper body weight of 400 N in Figure 9.4.1 changes to 379 N in this example.

	30-cm-High Belt	*70-cm-High Belt*
	S = 379 sin 44°	S = 379 sin 14°
	S = 263 N	S = 92 N
	C = 379 cos 44° + 3050	C = 379 cos 14° + 1815
	C = 3322 N	C = 2183 N

Answers to Discussion Questions in 9.3

1. Lifting work is performed when the conveyor belt is 30 cm high but not when the belt is rebuilt to a 70-cm height. In the latter case, only carrying work is performed because moving the object involves horizontal motion but no vertical motion. The number of lifts per day is included in this example so that the total amount of lifting work can be determined.

When changes in a work area are under consideration, it is essential to observe how the lifting work is performed in terms of the weight of the object, the vertical distance that the object is lifted, and the number of lifts performed per unit of time. These observations may reveal that certain changes may not affect the total amount of lifting work; i.e., the change may simply convert the work from a small number of heavy lifts to a large number of light lifts. Biomechanical and epidemiological data have not yet shown whether a greater risk of spinal injury results from many light lifts or from a few heavy lifts within the same time period.

Keyserling et al. (1980) conducted a biomechanical analysis of a work task in order to calculate the amount of lifting work required by the task and hence the erector spinae muscle strength needed by the worker. The study led to the design of four tests, which the job applicant was required to take before being hired. Over an 8-year period, significantly fewer spinal injuries occurred in employees who met the test requirement before being hired than in a control group of employees who were not tested. The study suggests, therefore, that it is efficient to adapt the amount of lifting work performed to the individual and to match the job requirements with the workers' lifting capability.

2. The three most important factors affecting the magnitude of the loads on the lumbar spine during lifting are:

- The distance from the lifted object to the center of motion in the lumbar spine (the length of the moment arm of the weight of the object).
- The length of the moment arm of the weight of the upper body.
- The weight of the lifted object.

The external torque on the lumbar spine can be reduced by changing one or more of these factors. The most efficient way to reduce this torque, and consequently the loads on the lumbar spine, is to minimize the length of the moment arm of the weight of the upper body by holding the lifted object as close to the body as possible (Andersson et al., 1977; Schultz et al., 1981).

Patients are often instructed to maintain a natural lordosis while lifting. However, the magnitude of the loads on the lumbar spine during lifting is not appreciably affected by the lordosis of the lumbar spine (Schultz, personal communication). It is possible that the distribution of loads on the lumbar spine is affected by the lordosis, however. (See also the discussion for example 9.2 and the commentary in this chapter.)

9.5. Commentary

A wealth of literature exists on the biomechanics of the lumbar spine. The reader wishing to pursue the subject in more depth is encouraged to read the following: Lindh, M., Biomechanics of the Lumbar Spine (Frankel and Nordin, 1980), and Panjabi, M., and White, A., Physical Properties and Functional Biomechanics of the Spine (White and Panjabi, 1978).

Loads on the Lumbar Spine

Biomechanical calculations of the loads on the lumbar spine have been validated experimentally by electromyography and by discometry (Andersson et al., 1977; Schultz et al., 1982b; Nachemson and Elfström, 1970). Loads on the lumbar spine were studied for over 250 different standing and sitting positions, with and without a lifted object, in healthy young test subjects. In these studies, free body diagrams were used to determine the main forces acting on the free body and the moment arms of these forces. From this information, the external torque on the lumbar spine was calculated. The erector spinae and abdominal muscle forces, as well as the compressive force on the L3 disc, were then predicted.

A mathematical model was used to predict the loads on the various tissues of the lumbar spine during these different sitting and standing positions (Schultz and Andersson, 1981). In the model, the computer was programmed to select the various muscles or muscle groups involved in maintaining a certain trunk position when a given external torque was produced on the lumbar spine.

The program was set up to select from among the muscles or muscle groups extending into the section of the body in which the muscle forces and the compressive force were acting. The principle on which this choice was structured was that the load on a given disc would be as low as possible and that no muscle force would exceed 100 N/cm^2. The program was devised to divide the muscle work among all muscles and muscle groups in the transverse body section under consideration. (For a description of this mathematical model, see Chapter 7, Section 7.10, Commentary.)

This method of calculating muscle activity has been validated by electromyographic studies of erector spinae and abdominal muscle activity. In these studies, the activity of the muscle group was recorded with surface electrodes. This method is satisfactory for recording symmetrical muscle activity but somewhat less satisfactory when the activity is asymmetrical.

The compressive force on the L3 disc calculated by means of the mathematical model was also validated through discometry. The disc pressure was measured by means of a miniature pressure transducer built into the tip of a needle that was inserted into the nucleus pulposus between the L3 and L4 vertebrae (Nachemson and Elfström, 1970). The signal from the pressure transducer was amplified and displayed on a recorder.

Comparisons of the three methods of study (biomechanical calculation, surface electromyography, and discometry) revealed excellent correlations for calculations of loads on the lumbar spine in the sagittal plane. Calculations of simultaneous forward-bending motion with rotation yielded good correlations for the three methods (Schultz et al., 1982a; Schultz et al., 1982c).

The lumbar discs are subjected to large forces in vivo. When an individual assumes an upright relaxed position, the compressive load on the lumbar discs corresponds approximately to the weight of the body. During sitting and standing in the forward-bending position when an object of 20 kg is lifted, the compressive force on the L3 disc increases to 5 times body weight. Walking, performing slight lateral bends, coughing, sneezing, or sitting with a lumbar spine support imposes a load of about 1.5 times body weight on the disc (Nachemson, 1963). The disc is weaker in torsion and in bending than in compression. Compression tests conducted on a motion segment in the lumbar spine did not lead to disc rupture but led to fractures in the vertebrae end plates (White and Panjabi, 1978).

Distribution of Loads on Various Structures of the Motion Segment

Although it is known that the facet joints are subjected to loads during motion, the magnitude of these loads during different types of motion has not been determined fully. In an in vitro study, Nachemson (1960) measured the disc pressure in the lumbar spine both in a normal motion segment and in one from which the facet joints had been removed. The study revealed that about 20% of the compressive load on the lumbar spine is absorbed by the facet joints. King et al. (1975), in another in vitro study, investigated the loads on the facet joints in a large number of specimens. The load on the facet joints varied between 0% and 33% of the total load on the lumbar spine and varied with changes in the position of the spine.

Farfan (1973) showed that the facet joints and disc in vitro (including the longitudinal posterior and anterior ligaments) each absorb about 45%, and together about 90%, of the torque around the motion segment.

These studies show that the disc and the vertebral body absorb most of the loads imposed on the lumbar spine. The facet joints absorb loads and exert a stabilizing function primarily during rotatory motion of the motion segments.

Assumptions Made in Problems 9.1 through 9.3

1. In the examples, the disc is assumed to be normal. Thus, the center of motion of every motion segment in the lumbar spine is placed in the disc. Rolander (1966) showed that when a motion segment in the lumbar spine has degenerated, the center of motion shifts to a point outside the disc. This finding can be disregarded in comparisons of the torques and forces acting on the lumbar spine of a single individual in two or more positions.

2. During relaxed upright standing, the inclination of the L3 disc is assumed to parallel the horizontal plane, and the angle of inclination of the L5 disc is 30° to this plane. Farfan (1973) found that during relaxed upright standing the proximal end plate of the L4 vertebra was parallel to the horizontal plane. Floyd and Silver (1955) found that for the same position the end plate of the sacrum is inclined 30° to the horizontal plane.

10

The Cervical Spine

The cervical spine is composed of the atlanto-occipital joint and seven motion segments. (The definition of a motion segment is given in the introduction to Chapter 9.) During flexion-extension of the atlanto-occipital joint, the axis of motion passes through the center of the mastoid processes (White and Panjabi, 1978). The transverse axis of motion for each motion segment in the cervical spine passes through the anterior portion of the subjacent vertebra (Lysell, 1969). In all of the examples in this chapter, motion of the cervical spine is considered in the sagittal plane.

Because traction of the cervical spine illustrates mechanical principles well, four examples (10.1–10.4) are devoted to this form of treatment. In these examples, the center of motion is located in the atlanto-occipital joint. In example 10.5, the loads on the C5 disc (i.e., the disc in the C5–C6 motion segment) are calculated for two sitting positions.

10.1. Loads through the Chin during Traction Treatment of the Cervical Spine

Resolution of Forces

Different types of halters can be used in the application of cervical traction. This example involves the use of a halter in which the angle between the chin strap and the neck strap is unaffected by a change in the direction of the traction force. Also, the halter is constructed so that the total traction force is distributed between the chin and the neck.

Problems

10.1A. A patient is being treated with cervical traction with the use of a soft, inelastic traction halter around his chin and neck. A horizontal traction force (T) of 100 N is applied to the halter. The component (N) of the traction force on the neck acts at a 45° angle to this force, and the component (C) on the chin acts at a 30° angle to this force (Figure 10.1A).

How large is the force on the chin (C)?

10.1B. The direction of the traction force (T) is changed so that the force acts at a 30° angle to the horizontal plane. The force on the chin

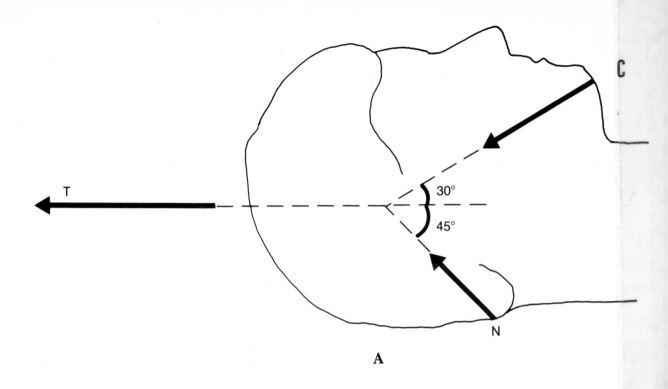

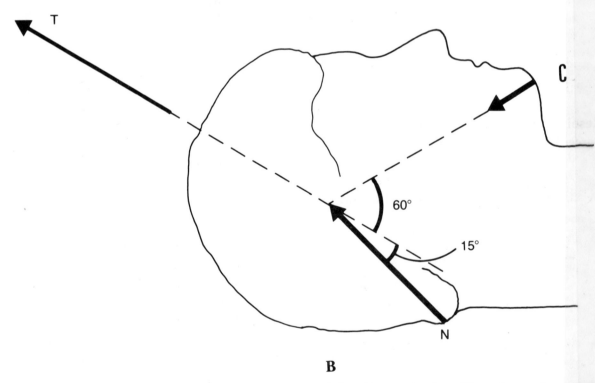

Figure 10.1. Cervical traction with a halter. The traction force (T) produces a force (C) on the chin and a force (N) on the neck. A. The angle between force T and force C is 30°, and between force T and force N, 45°. B.The angle between force T and force C is 60°, and between force T and force N, 15°.

(C) acts at a 60° angle to the traction force, and the force on the neck (N) acts at a 15° angle to the traction force (Figure 10.1B).

How large is the force on the chin (C)?

Discussion Questions

1. How large is the force in each chin strap in the two cases? Consider the traction halter as a linear structure. The chin is assumed to be a frictionless pulley, and the two chin straps (one on either side of the chin) are parallel.

2. The example shows that the direction of the traction force affects the distribution of force between the chin and the neck. Is there any other way to change the magnitudes of the forces on the chin and the neck without altering the magnitude of the traction force?

3. What joint is likely to be affected by the force on the chin?

10.2. Methods of Applying a Halter during Treatment with Cervical Traction

Torque

One indication for cervical traction is pain caused by compression of a nerve root (Shenkin, 1954; Stoddard, 1954). The aim of treatment is to stretch the soft tissues of the posterior cervical spine slightly and to enlarge the intervertebral foramina (Harris, 1977).

During normal neck flexion, the intervertebral foramina are enlarged, and the spinous processes separate (Lysell, 1969). Therefore, the importance of applying traction with the neck flexed is often stressed (Cailliet, 1964; Colachis and Strohm, 1965; Crue and Todd, 1965). The magnitude of the flexing torque depends on both the magnitude of the traction force and the length of its moment arms relative to the centers of motion in the atlanto-occipital joint and the cervical motion segments (Moritz, 1975).

This example shows that knowledge of the magnitude and direction of the traction force is not enough to permit an assessment of the mechanical effects of traction on the neck.

Problems

A supine patient is treated with cervical traction. The traction halter, which has an adjustable chin strap, is part of the Tru-trac system. The following four problems examine the torque on the atlanto-occipital joint produced by the traction force under four different traction conditions. In each case the traction force (T) is held constant at 120 N. The weight of the head and the friction between the head and the underlying support may be disregarded.

10.2A. In the first case, the traction force acts in a horizontal direction, and the chin strap is pulled tight (Figure 10.2A). How large is the

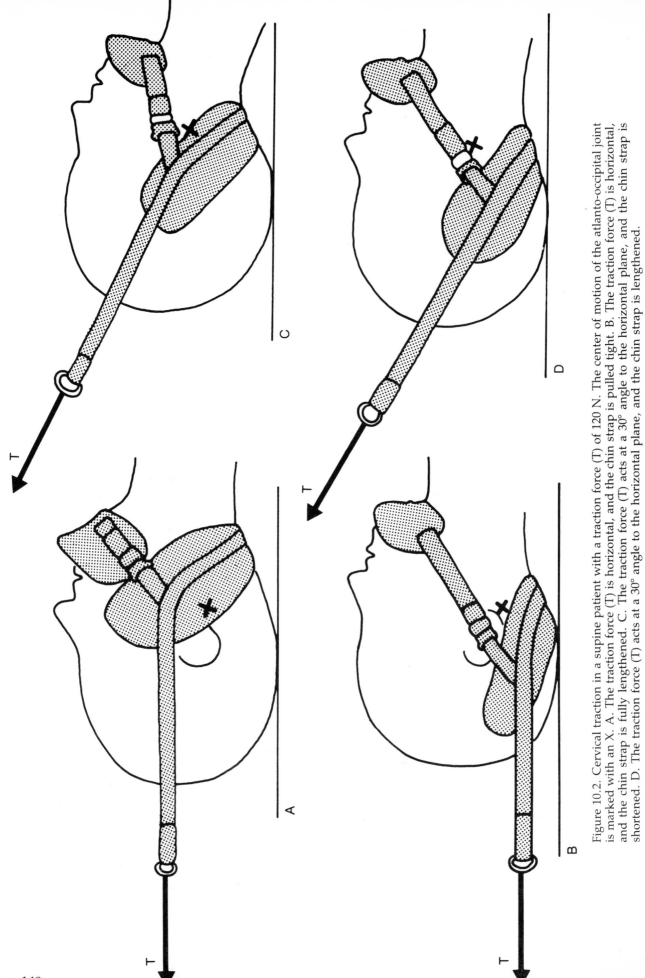

Figure 10.2. Cervical traction in a supine patient with a traction force (T) of 120 N. The center of motion of the atlanto-occipital joint is marked with an X. A. The traction force (T) is horizontal, and the chin strap is pulled tight. B. The traction force (T) is horizontal, and the chin strap is fully lengthened. C. The traction force (T) acts at a 30° angle to the horizontal plane, and the chin strap is shortened. D. The traction force (T) acts at a 30° angle to the horizontal plane, and the chin strap is lengthened.

torque (M) on the atlanto-occipital joint produced by the traction force if the moment arm is 0.04 m? Assume that the entire traction force is absorbed by this joint.

10.2B. In the second case, the chin strap is fully lengthened, but the direction of the traction force remains horizontal (Figure 10.2B). How large is the torque on the atlanto-occipital joint produced by the traction force if the moment arm is 0.02 m?

10.2C. In the third case, the direction of the traction force is altered so that the force acts at a 30° angle to the horizontal plane, and the chin strap is shortened (Figure 10.2C). How large is the torque on the atlanto-occipital joint produced by the traction force if the moment arm is 0.01 m?

10.2D. In the fourth case, the traction force acts at a 30° angle to the horizontal plane and the chin strap is lengthened (Figure 10.2D). How large is the torque on the atlanto-occipital joint produced by the traction force if the moment arm is 0.05 m?

Discussion Questions

1. What factors affect the magnitude of the torque in this example?
2. What mechanical factors determine the length of the moment arm of the traction force?

10.3. Analysis of Cervical Traction Using a Free Body Diagram

Moment Equilibrium; Force Equilibrium

This example provides experience in constructing a free body diagram and identifying the forces acting on the atlanto-occipital joint during traction treatment of the neck.

Problems

Traction is applied to the neck of a supine patient whose head is firmly supported. The patient is fully relaxed, and thus the forces acting on the head are absorbed by the atlanto-occipital joint. An arbitrary traction force (T) is applied to the head.

10.3A. Construct a free body diagram showing the forces acting on the free body. Include the friction force between the head and the underlying support. The center of gravity of the head and the arbitrary traction force (T) are shown in Figure 10.3, which is to be used in solving the problem.

10.3B. In Figure 10.3, draw in the moment arms of the pertinent forces relative to the center of motion of the atlanto-occipital joint. Devise the equations for the forces and the moments acting on the free body for both equilibrium conditions.

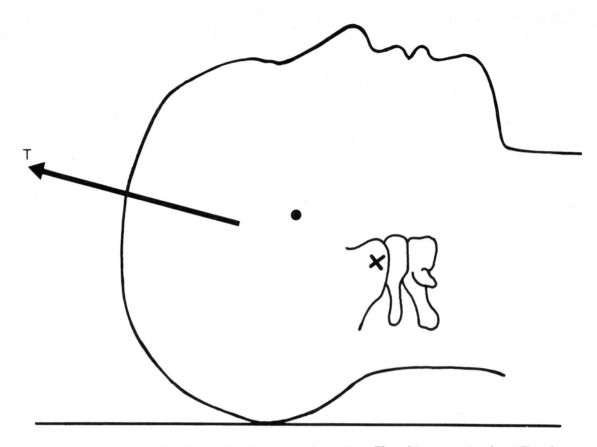

Figure 10.3. Cervical traction in a supine patient. The arbitrary traction force (T) is shown. The center of gravity of the head is marked with a solid dot, and the center of motion of the atlanto-occipital joint is marked with an X.

Discussion Question

Describe the mechanical effect of the friction between the head and the underlying support on the atlanto-occipital joint.

10.4. Loads on the Atlanto-occipital Joint during Treatment with Cervical Traction

Moment Equilibrium; Force Equilibrium

Traction is used to stabilize the cervical spine after injury. In this situation, the forces on the vertebral bodies produced by traction should be as axial as possible to minimize the bending moments and the shear forces acting on the cervical spine.

This example examines how large a traction force must be applied, and in what direction, for both the torque and the shear force at an arbitrary point in the cervical spine to equal zero. In the example, the center of motion of the atlanto-occipital joint is designated as the arbitrary point because flexion and extension of the neck begin in this joint.

The free body diagram from example 10.3 may be useful in solving the following problems. For definitions of the sine, cosine, and tangent of an angle, refer to Chapter 1, Section 1.2, Useful Formulas and Concepts.

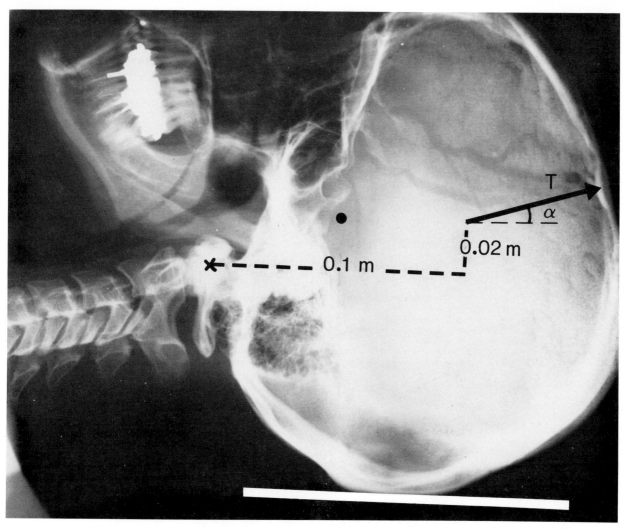

Figure 10.4. Cervical traction in a supine patient. A traction force (T) is applied 0.1 m cranial to and 0.02 m ventral to the center of motion of the atlanto-occipital joint (marked with an X) in the direction α relative to the horizontal plane. The center of gravity of the head is marked with a solid dot.

Problems

A patient with an injury to the cervical spine is treated with traction in the supine position.

10.4A. How large a traction force (T) must be applied, and in what direction α relative to the horizontal plane must it act, for there to be no torque and no shear force on the atlanto-occipital joint? The traction force (T) is applied 0.1 m cranial to and 0.02 m ventral to the center of motion of the atlanto-occipital joint (Figure 10.4).

The normal force between the head and the underlying support is not changed by the application of traction. Because the patient is relaxed, the forces acting on the head are absorbed by the atlanto-occipital joint. The joint surfaces are assumed to parallel the vertical plane.

The weight of the head is 50 N, and the center of gravity is 0.05 m cranial to the center of motion of the atlanto-occipital joint.

The force of the head on the underlying support (the normal force) is 36 N, and the contact point is 0.07 m cranial to and 0.08 m dorsal to the center of motion. The coefficient of friction between the head and the underlying support is 0.2.

10.4B. How large is the tensile force (V) on the atlanto-occipital joint? Necessary data can be obtained from the preceding problem.

Discussion Questions

1. Is the point of application of the traction force important to the conditions set forth in this example, namely, that no torque and no shear forces act on the atlanto-occipital joint? If the traction force (T) is applied 0.1 m cranial to and 0.01 m dorsal to the atlanto-occipital joint, how large must this force be, and in what direction (α) must it act, for the torque and the shear forces on the atlanto-occipital joint to equal zero?

2. What factors affect the magnitude of the tensile force (V) on the atlanto-occipital joint? Assume that the patient is relaxed and that the applied force is absorbed by the atlanto-occipital joint.

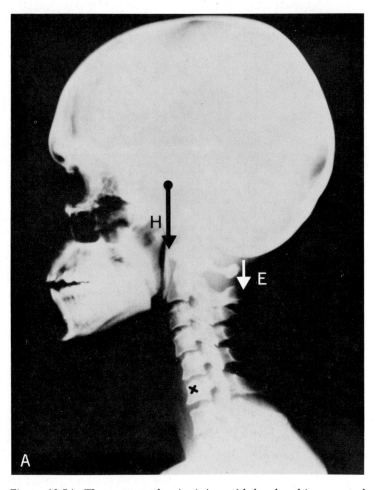

Figure 10.5A. The cartographer is sitting with her head in a neutral position. The center of motion of the C5–C6 motion segment is marked with an X. The weight of the head (H) is counteracted by the erector spinae muscle force (E). (From Svedenkrans, 1980.)

10.5. The Reaction Force on the C5 Disc with the Head in a Neutral and a Flexed Position

This example shows how the reaction force on the disc can be calculated when the magnitude of the muscle force is unknown. In the example, the reaction force on the C5–C6 motion segment when an individual holds her head in a neutral position is compared with that when the head is bent forward. The magnitude of the reaction force in the C5–C6 motion segment is of interest because degenerative changes in the cervical spine most often occur in this region (Friedenberg and Miller, 1963).

Problems

A cartographer is seated at her drawing board. The weight of her head (H) is 40 N. The forward-bending moment in the C5–C6 motion segment, produced by the weight of the head, is counteracted by the erector spinae muscle force (E).

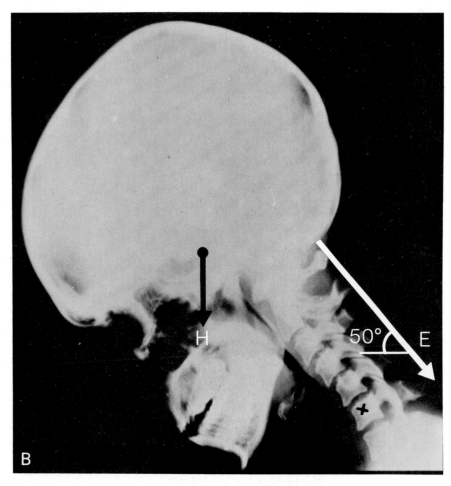

Figure 10.5B. The cartographer is sitting with her head flexed. The center of motion of the C5–C6 motion segment is marked with an X. The weight of the head (H) is counteracted by the erector spinae muscle force (E), whose line of application forms an angle of 50° with the horizontal plane. (From Svedenkrans, 1980.)

10.5A. How large is the reaction force (R) on the C5 disc when the cartographer sits with her head in a neutral position (Figure 10.5A)? The moment arm of the weight of the head is 0.02 m. The erector spinae muscle force (E) acts in a vertical direction, and its moment arm is 0.04 m (Svedenkrans, 1980).

10.5B. The cartographer flexes her head. How large is the reaction force (R) on the C5 disc if the erector spinae muscle force (E) acts at an angle of 50° to the horizontal plane? Necessary data can be obtained from Figure 10.5B.

10.6. Answers, Solutions, and Discussion

Answers to the Problems

10.1A. $C = 73 \pm 2N$

10.1B. $C = 27 \pm 2N$

10.2A. 4.8 Nm (extending torque)

10.2B. 2.4 Nm (flexing torque)

10.2C. 1.2 Nm (flexing torque)

10.2D. 6.0 Nm (flexing torque)

10.3B. The clockwise moments are positive. The forces directed vertically upward and horizontally to the left are positive. For an explanation of the symbols in the following equations, see the solutions to problems 10.3A and B.

$\Sigma M = 0$ gives $t_1T \sin \alpha - t_2T \cos \alpha - hH + nN - fF = 0$

$\Sigma F_y = 0$ gives $T \sin \alpha - H + N - S = 0$

$\Sigma F_x = 0$ gives $T \cos \alpha - F - V = 0$

10.4A. $T = 44$ N
   $\alpha = 18°$

10.4B. $V = 35$ N

10.5A. $R = 60$ N

10.5B. $R = 148 \pm 25$ N

The ranges shown for some of these answers allow for a reasonable margin of error in the graphic solution of the problems. These ranges were determined on the basis that 10 N corresponds to 1 cm and that the maximum allowable error in measurement is 1 mm of length and 1° of angle.

Solution to Problem 10.1A

10.1A. Find the force (C) on the chin of a patient being treated with cervical traction.

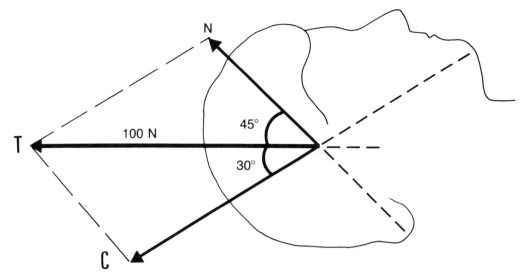

Figure 10.6.1. Cervical traction with a halter. The traction force (T) is 100 N. The force on the chin (C) and the force on the neck (N) are known to act at angles of 30° and 45°, respectively, to the traction force (T). The magnitude of force C is obtained by means of the parallelogram method.

Graphic Solution

Force T represents the resultant of force components N and C. A parallelogram of forces and the theorem of displacement are used to obtain the magnitude of these components.

The intersection point of forces T, N, and C is located. Force T is moved along its line of application to this intersection point. The lines of application of force N and force C are then extended.

A parallelogram of forces is constructed with T as the diagonal and with N and C as the sides (Figure 10.6.1). The side representing force C is measured.

$$C = 73 \pm 2 \text{ N}$$

Mathematical Solution

Force C is obtained from Figure 10.6.1 by means of the sine theorem.

$$\frac{C}{\sin 45°} = \frac{100}{\sin (108° - [30° + 45°])}$$

180

$$C = \frac{100 \sin 45°}{\sin 105°}$$

$$C = 73 \text{ N}$$

Answers to Discussion Questions in 10.1

1. The force in each chin strap is half as large as the force on the chin, i.e., about 35 N in problem 10.1A and about 15 N in problem 10.1B (Figure 10.6.2). The line of application of the traction force corresponds to the alignment of the traction. The chin acts as a pulley, and the straps on either side of the chin are parallel. The force on the chin is the resultant of the forces in the two chin straps.

2. Provided that the angle between the force on the chin (C) and the force on the neck (N) does not change, the force on the chin decreases

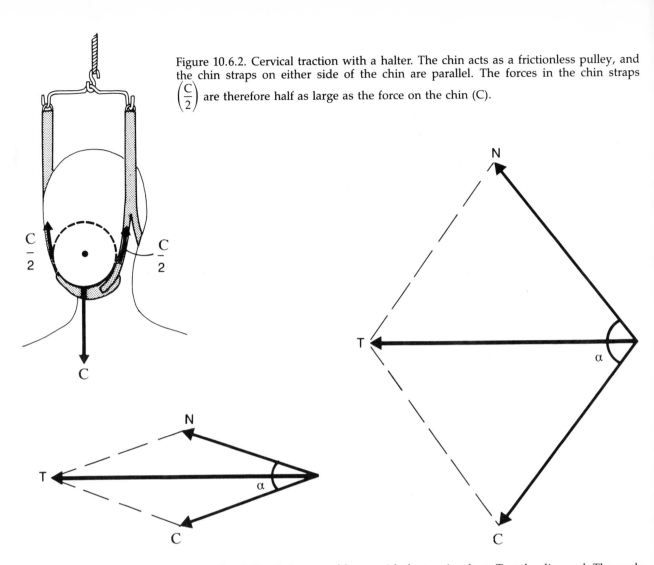

Figure 10.6.2. Cervical traction with a halter. The chin acts as a frictionless pulley, and the chin straps on either side of the chin are parallel. The forces in the chin straps $\left(\dfrac{C}{2}\right)$ are therefore half as large as the force on the chin (C).

Figure 10.6.3. Parallelogram of forces with the traction force T as the diagonal. The angle α between the force components N and C is smaller in the left figure, and thus the magnitudes of N and C are lower.

when the direction of the traction force (T) is shifted (i.e., angled upward). The size of the angle (α) between force C and force N also affects the magnitude of these components. The larger the angle is, the greater the magnitude of the force components if the magnitude of force T is held constant (Figure 10.6.3). In practice this means that the angle between force C and force N should be kept small to minimize the loads on the chin and neck. When the traction halter has an adjustable chin strap, the angle is minimized by lengthening the strap (see example 10.2).

3. The application of cervical traction can produce loads on the temporomandibular joint. Frankel and Burstein (1970) found that long-term loading of this joint causes pain. Therefore, they recommended that a patient with an uneven bite use a bite plate when treated with cervical traction. Use of the bite plate creates a more even distribution of the loads on the temporomandibular joint. As an alternative to the use of a bite plate in some cases, the traction rope can be angled upward and/or the angle between the chin and neck straps can be kept to a minimum.

Solution to Problem 10.2A

Find the torque (M) on the atlanto-occipital joint produced by the traction force.

$M = 120 \cdot 0.04$

$M = 4.8 \ Nm$

An extending torque is exerted on the neck (Figure 10.6.4).

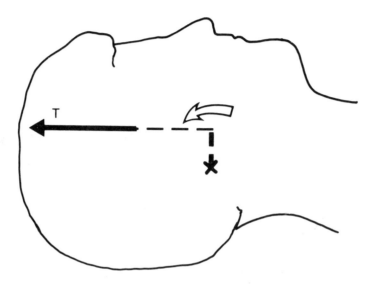

Figure 10.6.4. Cervical traction in a supine patient. The traction force (T) acts ventral to the center of motion of the atlanto-occipital joint and thus produces an extending torque on the neck.

Answers to Discussion Questions in 10.2

1. When a traction halter with an adjustable chin strap is used, the magnitude of the torque is affected by the amount the chin strap has been tightened and by the direction of the traction force.

2. Theoretically, the length of the chin strap determines the point of application of the traction force (T). The length of the moment arm of the traction force (T) relative to the center of motion of the atlanto-occipital joint is determined by the point of application and direction of force T.

In calculations of the magnitude of the torque on the atlanto-occipital joint in the sagittal plane, the position of the point of application of the traction force T in the ventrodorsal direction and in the craniocaudal direction affects the moment arm of this force. The significance of the position of the point of application in the ventrodorsal direction is illustrated in problems 10.2A and B and in Figures 10.6.4 and 10.6.5. If the line of application of the traction force (T) passes ventral to the center of motion of the atlanto-occipital joint, the force acts to extend the neck (Figure 10.6.4); if it passes dorsal to the center of motion, it produces a flexing torque on the neck (Figure 10.6.5). The significance of the position of the point of application in the craniocaudal direction is illustrated in Figure 10.6.6 and by problems 10.2C and D. If the magnitude and direction of the traction force (T) are held constant, the moment arm lengthens as the point of application of the force is shifted cranially, and thus the flexing torque increases (Figure 10.6.6). Thus, both the type of halter used and the positioning of the halter affect the magnitude of the torque.

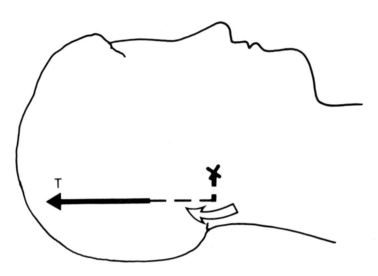

Figure 10.6.5. Cervical traction in a supine patient. The traction force (T) acts dorsal to the center of motion of the atlanto-occipital joint and thus produces a flexing torque on the neck.

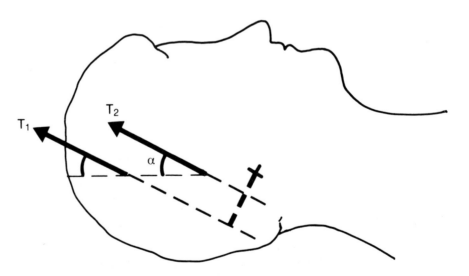

Figure 10.6.6. Cervical traction in a supine patient. Two traction forces (T_1 and T_2) with equal magnitudes and the same direction (α) with respect to the horizontal plane but different points of application produce torques of different magnitudes on the atlanto-occipital joint. T_1 acts cranially with respect to T_2 and thus produces a larger flexing torque on the neck.

For cervical traction treatment to be standardized, two characteristics of the traction force must be noted each time traction is applied:

- The magnitude of the traction force.
- The line of application (angle of rope pull) relative to a surface landmark associated with a center of motion (for example, the center of motion of the atlanto-occipital joint).

Both characteristics must be known for a complete biomechanical analysis of the forces and moments acting on the cervical spine to be performed.

For the neck to flex in a supine patient being treated with cervical traction, the flexing torque produced by the traction force must exceed the extending torque produced by the weight of the head, which is 1–2 Nm (not considered in this problem). If the traction force is about 100 N, its line of application must pass about 1 cm dorsal to the mastoid processes for the extending torque to be overcome.

Solutions to Problems 10.3A and B

10.3A. A free body diagram is constructed of the head of a supine patient being treated with cervical traction showing the forces acting on

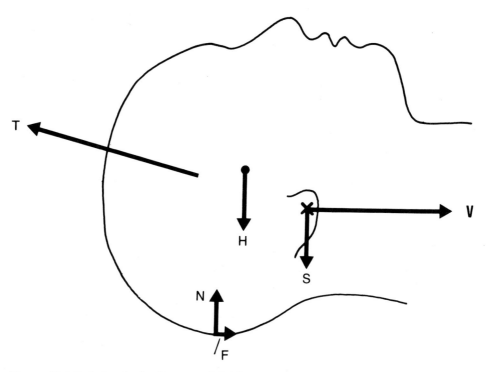

Figure 10.6.7. A free body diagram of the forces acting on the head of a supine patient being treated with cervical traction. The head is firmly supported. The center of motion of the atlanto-occipital joint is marked with an X.

T = the traction force.
H = the weight of the head.
N = the force of the head on the underlying support.
F = the friction force between the head and the underlying support.
V = the horizontal component of the atlanto-occipital joint reaction force.
S = the vertical component of the atlanto-occipital joint reaction force.

the head. It is assumed that the forces acting on the head are counteracted by the reaction force on the atlanto-occipital joint. Because the magnitude and direction of the reaction force are unknown, this force is divided into a vertical component (S) and a horizontal component (V). The following forces act on the free body (Figure 10.6.7):

The traction force (T)

The weight of the head (H)

The force of the head on the underlying support (N)

The friction force between the head and the underlying support (F)

The horizontal component of the atlanto-occipital joint reaction force (V)

The vertical component of the joint reaction force (S)

10.3B. Equilibrium equations are devised for the forces shown on the free body diagram in problem 10.3A.

The traction force T acts at an angle of α to the horizontal plane.

The vertical force component of T is $T \sin \alpha$.

The horizontal force component of T is $T \cos \alpha$ (Figure 10.6.8).

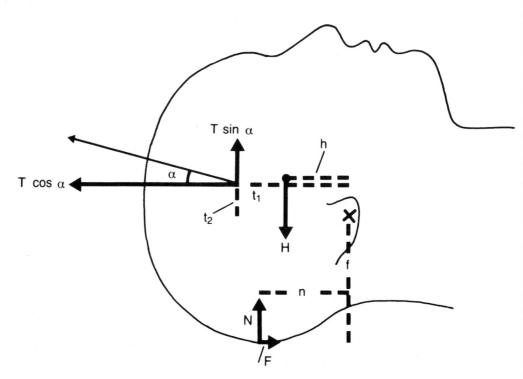

Figure 10.6.8. Cervical traction in a supine patient. The forces acting on the head and their moment arms relative to the center of motion of the atlanto-occipital joint are indicated. The vertical force component ($T \sin \alpha$) of the traction force (T) has the moment arm t_1, and its horizontal force component ($T \cos \alpha$) has the moment arm t_2. The moment arm of the weight of the head (H) is h, the moment arm of the force of the head on the underlying support (N) is n, and that of the friction force between the head and the underlying support (F) is f.

The following moment arms are present (Figure 10.6.8):

t_1 of force $T \sin \alpha$
t_2 of force $T \cos \alpha$
h of force H
n of force N
f of force F

The condition for static equilibrium yields three equations.

The clockwise moments are positive. The forces directed vertically upward and horizontally to the left are designated as positive.

$$\Sigma M = 0 \text{ gives } t_1 T \sin \alpha - t_2 T \cos \alpha - hH + nN - fF = 0 \tag{1}$$

$$\Sigma F_y = 0 \text{ gives } T \sin \alpha - H + N - S = 0 \tag{2}$$

$$\Sigma F_x = 0 \text{ gives } T \cos \alpha - F - V = 0 \tag{3}$$

Answer to Discussion Question in 10.3

The friction between the head and the underlying support causes a portion of the traction force (T) to "dissipate" and reduces the tensile force (V) on the cervical spine (see the solution to problem 10.3B, equation 3). The friction force also produces an extending moment on the atlanto-occipital joint (see the solution to problem 10.3B, equation 1). This force can be decreased either by reducing the coefficient of friction (for example, by means of a gliding board) or by decreasing the force of the head on the underlying support.

Solutions to Problems 10.4A and B

In these problems, three conditions are assumed: (1) there is no torque acting on the atlanto-occipital joint ($\Sigma M = 0$); (2) there is no shear force on this joint; and (3) the normal force produced by the weight of the head on the underlying surface (36 N) remains unchanged during traction treatment.

10.4A. Find the magnitude of the traction force (T) and its direction (α) relative to the horizontal plane in a supine patient being treated with cervical traction.

The following forces act on the head (Figure 10.6.9):

The weight of the head (50 N), acting vertically downward.

The weight of the head on the underlying surface (36 N), acting vertically upward.

The friction force between the head and the underlying support ($0.2 \cdot 36 = 7.2$ N), acting horizontally to the left.

The traction force (T), acting in the direction α relative to the horizontal plane.

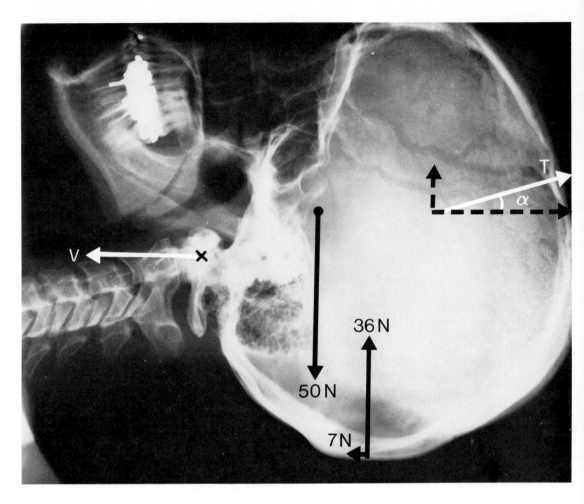

Figure 10.6.9. Free body diagram of the head of a supine patient being treated with cervical traction. The weight of the head (50 N) produces a normal force of 36 N on the underlying support. The friction force between the head and the underlying support is 7 N. The traction force (T) is such that only a tensile force (V) is exerted on the atlanto-occipital joint. The center of motion of this joint is marked with an X.

The tensile force (V). Because there is no shear force, this force is perpendicular to the atlanto-occipital joint surface and thus acts horizontally to the left.

In this problem, the clockwise moments are designated as negative.

$\Sigma M = 0$ gives the relationship between the force components T sin α and T cos α.

0.1 T sin α − 0.02 T cos α − 0.05 · 50 + 0.07 · 36 − 0.08 · 7.2 = 0

0.1 T sin α − 0.02 T cos α = 0.56   (1)

The forces directed vertically upward are positive.

$\Sigma F_y = 0$ gives the magnitude of force T sin α.

T sin α − 50 + 36 = 0

T sin α = 14   (2)

Combining equation 2 and equation 1 gives the magnitude of force T cos α.

$0.1 \cdot 14 - 0.02 \: T \cos \alpha = 0.56$

$0.02 \: T \cos \alpha = 1.4 - 0.56$

$T \cos \alpha = 42$ (3)

The result of equation 2 is divided by the result of equation 3.

$$\frac{T \sin \alpha}{T \cos \alpha} = \frac{14}{42}$$

T can be canceled out. Because $\dfrac{\sin \alpha}{\cos \alpha} = \tan \alpha$, the equation can now be written:

$$\tan \alpha = \frac{14}{42}$$

$\tan \alpha = 0.333$

$\alpha = 18.4°$

$\alpha = 18.4°$ is substituted into equation 2.

$T \sin 18.4° = 14$

$$T = \frac{14}{\sin 18.4°}$$

$T = 44 \: N$

$\alpha = 18°$

10.4B. Find the tensile force (V) on the atlanto-occipital joint in the same patient as in the preceding problem.

Three forces are considered (see Figure 10.6.9): force T cos α, force V, and the friction force (rounded off to 7 N). The solution to problem 10.4A gives the magnitude of the traction force (T = 44 N), the direction of the force relative to the horizontal plane (α = 18°), and the friction force between the head and the underlying support (7 N).

In this problem, the horizontal forces directed to the right are designated as positive.

$\Sigma F_x = 0$ gives the magnitude of the tensile force V.

$44 \cos 18° - 7 - V = 0$

$V = 35 \: N$

Answers to Discussion Questions in 10.4

1. In this example, the point of application of the traction force does affect the conditions set forth, namely, that no torque and no shear forces are produced on the atlanto-occipital joint. If the traction force (T) is applied 0.01 m dorsally with respect to the center of motion instead of

0.02 m ventrally as in the example, the problem has no solution because, with this point of application of the traction force, the shear forces and the torque on the atlanto-occipital joint cannot both equal zero.

For a torque to be prevented when the head rests on the underlying support, the line of application of the traction force must pass through the mastoid processes. For shear forces to be prevented, in this example the point of application of the traction force must lie somewhat ventral to the mastoid processes so that the vertical component of force T counterbalances the difference between the weight of the head and the force of the head on the underlying support.

If the traction force (T) is applied dorsally with respect to the center of motion, the problem can be solved as follows.

In this problem, the clockwise moments are designated as negative.

$\Sigma M = 0$ gives the relationship between the force components $T \sin \alpha$ and $T \cos \alpha$.

$$0.1\ T \sin \alpha + 0.01\ T \cos \alpha - 0.05 \cdot 50 + 0.07 \cdot 36 - 0.08 \cdot 7.2 = 0$$

$$0.1\ T \sin \alpha + 0.01\ T \cos \alpha = 0.556 \tag{1}$$

The forces directed vertically upward are positive.

$\Sigma F_y = 0$ gives the force component $T \sin \alpha$.

$$T \sin \alpha - 50 + 36 = 0$$

$$T \sin \alpha = 14 \tag{2}$$

The result of equation 2 is substituted into equation 1:

$$0.1 \cdot 14 + 0.01\ T \cos \alpha = 0.556 \tag{3}$$

$$0.01\ T \cos \alpha = -0.84$$

$$T \cos \alpha = -84$$

The result of equation 1 is divided by the result of equation 3:

$$\frac{T \sin \alpha}{T \cos \alpha} = \frac{-14}{84}$$

$$\tan \alpha = -0.1667$$

$$\alpha = -9.5°$$

$$T = -84.8\ N$$

The negative value for α means that the line of application of the force lies 9.5° below the horizontal plane. The negative value for force T means that the force is directed to the left against the atlanto-occipital joint and thus does not produce traction but works compressively instead. With this point of application for force T, traction invariably produces a torque and/or shear forces on the atlanto-occipital joint.

2. The magnitude of the tensile force is dependent on the magnitude and direction of the traction force and on the magnitude of the friction force between the head and the underlying support (see the solution to problem 10.4B). Furthermore, the point of application of the traction force

may affect the distribution of the loads on the atlanto-occipital joint. For example, if a torque is produced on this joint, the largest tensile stress occurs dorsal to the center of motion, and a smaller tensile stress—or even a compressive stress—is produced ventral to the center of motion.

Solutions to Problems 10.5A and B

10.5A. Find the reaction force (R) on the C5 disc when the cartographer is sitting with her head in a neutral position.

The following forces act on the head: the weight of the head (H) (40 N) acting vertically downward, the erector spinae muscle force (E) acting vertically downward, and the reaction force (R) on the C5 disc.

In this problem, the clockwise moment is positive.

$\Sigma M = 0$ gives the magnitude of the erector spinae muscle force E.

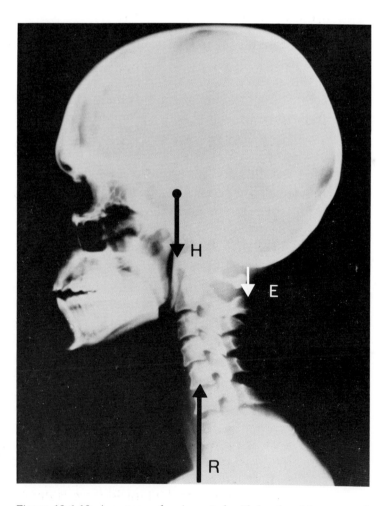

Figure 10.6.10. A cartographer is seated with her head in a neutral position. The weight of the head (H) and the erector spinae muscle force (E) produce a vertical reaction force (R) on the C5 disc.

165

$$0.04\ E - 0.02 \cdot 40 = 0$$

$$E = 20\ N$$

The forces directed vertically downward are positive.

$\Sigma F_y = 0$ gives the magnitude and direction of the reaction force R on the C5 disc.

$$R + 20 + 40 = 0$$

$R = -60\ N$ (i.e., directed upward) (Figure 10.6.10)

10.5B. Find the reaction force (R) on the C5 disc when the cartographer flexes her head.

The following forces act on the head: the weight of the head (H) (40 N) acting vertically downward, the erector spinae muscle force (E) acting at a 50° angle to the horizontal plane, and the reaction force (R) on the C5 disc acting in direction α relative to the horizontal plane.

Graphic Solution

The magnitude of the reaction force R is determined in the following manner.

When three nonparallel forces act on a body in equilibrium, all characteristics of the forces can be obtained if the points of application for all three forces, the directions for two forces, and the magnitude of one force are known. Because the body is in equilibrium, the lines of application of the three forces acting on the body must intersect at one point. Therefore, the direction of the third force can be determined. Once the directions of all three forces are known, a polygon of forces can be constructed from which the magnitudes of two forces can be scaled.

In this problem, force H is fully defined, and the points of application of E and R are known (R must pass through the center of motion of the atlanto-occipital joint in order not to exert a moment on the joint). The direction of force E is given. The direction (α) of the reaction force (R) on the C5 disc is obtained as follows from Figure 10.6.11. The lines of application of force H and force E meet at a point. Force R must also pass through this point. Because two points through which force R passes are now known (the point of intersection of forces H and E and the center of motion), the line of application of force R can now be determined. The direction α is measured and found to be 60° to the horizontal plane.

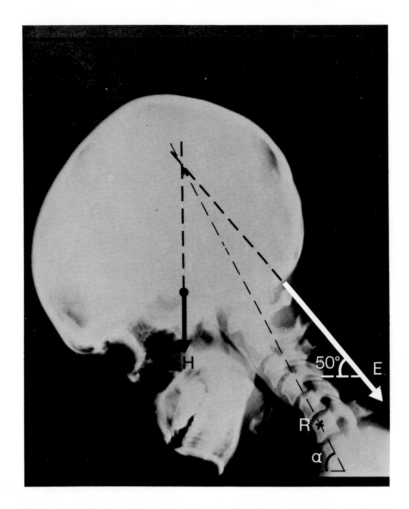

Figure 10.6.11. A cartographer is seated with the head flexed. The lines of application of the weight of the head (H) and the erector spinae muscle force (E) intersect at one point. With the head in equilibrium, the line of application of the reaction force (R) on the C5 disc must also pass through this point. Force E forms a 50° angle, and force R an angle of α°, with the horizontal plane.

The magnitude of force R is then obtained by means of the polygon method (Figure 10.6.12). The lines of application for forces E and R are extended until they intersect. The distance from this intersection point to the point of intersection of R and H gives the magnitude of force R.

R = 148 ± 25 N

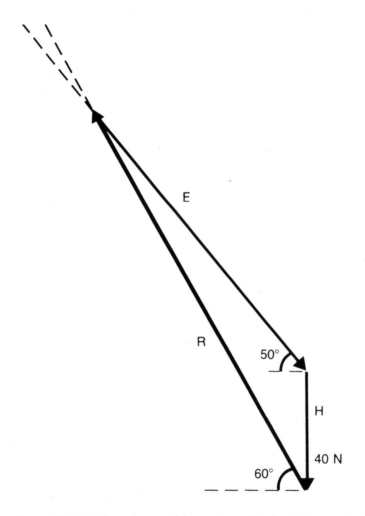

Figure 10.6.12. The polygon of forces is used to find the magnitude of the reaction force (R) on the C5 disc. The weight of the head (H = 40 N) acts vertically downward. The erector spinae muscle force (E) forms a 50° angle with the horizontal plane. The reaction force (R) on the C5 disc forms a 60° angle with this plane.

The direction of force R (60°) is measured with a protractor in Figure 10.6.11. Forces E and R are resolved into horizontal and vertical force components. The horizontal forces directed to the right and the vertical forces directed downward are positive.

$\Sigma F_x = 0$ gives the relationship between the erector spinae muscle force E and the reaction force R on the C5 disc. This relationship applies to the horizontal components of these forces.

$E \cos 50° - R \cos 60° = 0$

$$E = \frac{R \cos 60°}{\cos 50°}$$

$E = 0.78\ R$

$\Sigma F_y = 0$ gives another relationship between the erector spinae muscle force E and the reaction force R on the C5 disc. This relationship applies to the vertical components of these forces.

$E \sin 50° - R \sin 60° + 40 = 0$

$E = 0.78\ R$ is substituted into this equation.

$0.78\ R \sin 50° - R \sin 60° + 40 = 0$

$0.60\ R - 0.87\ R + 40 = 0$

$0.27\ R = 40$

$R = 148\ N$

10.7. Commentary

Cervical traction is used both as a treatment for conditions causing cervical pain and as a means of stabilizing fractures of the cervical vertebrae. The principle of traction treatment for cervical pain is that a forward-bending moment acts in conjunction with a tensile force on the cervical spine. The object of treatment is to produce an optimum increase in the size of the intervertebral foramina. Therefore, for various traction methods to be compared, not only must the magnitude and direction of the traction force be described, but also the line of application of this force relative to the center of motion of the joint under consideration (the moment arm). Unless the length of the moment arm of the traction force relative to the center of motion is known, the forward-bending moment on the joint cannot be calculated (see example 10.2).

During traction treatment, the forward-bending moment on the cervical spine, acting in conjunction with the tensile force, affects the cervical spine mechanically in several ways: the cervical lordosis is reduced (Shenkin, 1954), the facets are distracted (Bard and Jones, 1964), the intervertebral foramina are enlarged (Bard and Jones, 1964), and the vertebral bodies are distracted. (The distraction is greater on the dorsal sides of the vertebral bodies than on the ventral sides [Colachis and Strohm,

1966]). These mechanical effects in the cervical spine are influenced by the frequency and duration of traction treatment as well as by the amount of friction between the head and the underlying support. The position of the patient (e.g., supine or sitting) can also have a significant effect.

In the cervical spine, the passive structures are so rigid that very little axial motion takes place in each individual motion segment. In a study of axial cervical traction in eight healthy test subjects ranging in age from 18 to 37 years (Schlicke et al., 1979), a traction force equal to one-third of the body weight was applied for three minutes. The average amount of distraction between vertebral bodies was 0.7 ± 0.5 mm. The measurements involved the C2 to C6 vertebrae.

With the cervical spine in traction, the space between vertebral bodies may increase in one or more motion segments while remaining unchanged (or even diminishing) in others. The amount of distraction depends, among other things, on the degree of elasticity of the passive structures. Degenerative changes in a motion segment do not necessarily lead to increased distraction in the motion segment (Goldie and Reichman, 1977).

Appendix A

Body Segment Parameters

Data on the weight of individual body segments and the location of the centers of gravity given in the examples were obtained from the literature (Drillis et al., 1964; Walker et al., 1973) and are based on data from Dempster (1955). Table A–1 gives the weight of the body segments as a percentage of the entire body weight, Table A–2 lists the surface landmarks associated with anatomic centers and axes of motion of various joints, and Figure A–1 shows the location of the centers of gravity of various body segments as a percentage of the length of the body segment. The center of gravity of the head in the sagittal plane is located 2 cm ventral to and 1 cm cranial to the outer auditory canal (Aldman, 1981).

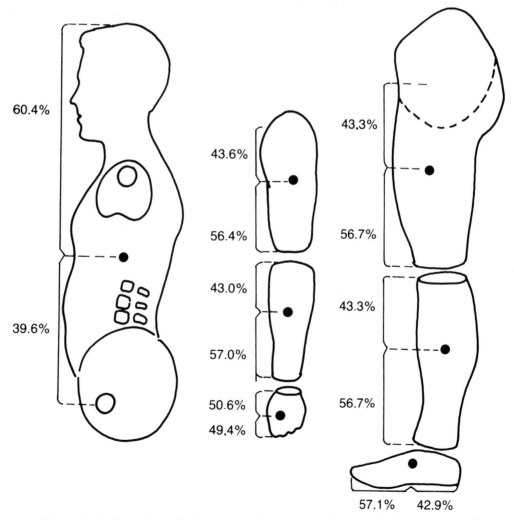

Figure A–1. Locations of the centers of gravity in the sagittal plane expressed as a percentage of the length of the body segments. (From Dempster, 1955.)

Table A–1. Weight of Body Segments as a Percentage of Total Body Weight

Body Segment	Percentage of Total Body Weight
Head and neck	7.9
Trunk with head and neck	56.5
Upper arm	2.7
Forearm	1.5
Hand	0.6
Thigh	9.7
Lower leg	4.5
Foot	1.4

Table A–2. Surface Landmarks Associated with Anatomic Centers and Axes of Motion of the Joints*

Shoulder	A point at the center of the palpable parts of the humeral head.
Elbow	Sagittal plane: The intersection point of a line between the lowest palpable point of the medial epicondyle of the humerus with a point 8 mm proximal to the humeroradial junction.
Hip	Sagittal plane: A point 1 cm anterior to the most laterally projecting part of the greater trochanter.
	Frontal plane: A point 2 cm distal to the midpoint of a line from the antero-superior iliac spine to the pubic symphysis (Andriacchi et al., 1981).
Knee	Sagittal plane: A line between the posterior convexities of the femoral condyles.
Ankle	Sagittal plane: The insection point of a line between the tip of the lateral malleolus of the fibula with a point 5 mm distal to the tibial malleolus.
Atlanto-occipital joint	Sagittal plane: A line through the centers of the mastoid processes (White and Panjabi, 1978).
Lumbar spine (L3–L4 motion segment)	Sagittal plane: A point 1–2 cm dorsal to the superior-most point on the iliac crest.
	Frontal plane: A point between the spinous processes of the L3–L4 vertebrae.

*If not otherwise indicated, the data are obtained from Williams and Lissner (1977).

Appendix B

The International System of Units (SI)

The International System of Units (SI) is used for all units of measure throughout this book. Table B–1 lists common SI units, and Table B–2 gives some British-metric equivalents.

Table B–1. Units of the International System of Units (SI)

Physical Quantity	Unit	Symbol	Relation to Other Units
Mass	kilogram	kg	
Length	meter	m	
Force	newton	N	$kg\ m/s^2$
Torque	newton-meter	Nm	
Energy, work	joule	J	Nm
Power	watt	W	J/s
Pressure and mechanical stress	pascal	Pa	N/m^2

Table B–2. Some British Metric Equivalents

1 kp = 9.81 N	1 N = 0.102 kp
1 lb = 4.448 N	1 N = 0.2248 lb
1 lb (weight) = 0.4536 kg (mass)	1 kg (mass) = 2.205 lb (weight)
1 in = 2.54 cm	1 cm = 0.3937 in
1 ft = 30.48 cm	1 m = 39.37 in
1 ft lbf = 1.356 J	1 J = 0.7376 ft lbf

Appendix C

Maximum Isometric Torque at Various Joint Angles

It is useful to know how the maximum torque normally changes through the range of joint motion when a muscle or muscle group is exercised. Figures C–1–C–8 show the maximum isometric torque at various joint angles for a number of muscle groups during isometric muscle strength tests. In the graphs, the maximum torque is presented on a relative scale (i.e., as a percentage of the highest value achieved for the torque over the range of joint motion). The joint angles are defined according to the anatomic starting positions described in *Joint Motion. Method of Measuring and Recording* (British Orthopaedic Association, 1965).

The shape of the curves obtained during isometric strength tests correspond in large measure to the shape of the curves obtained through isokinetic test methods. (The isokinetic strength is the maximum dynamic strength of a muscle or muscle group when exercised at a certain constant speed.) However, during isometric muscle testing, the torque developed is larger at every joint angle than during isokinetic muscle testing (Figure C–9). The faster an isokinetic exercise is performed, the lower the torque at every joint angle. Nevertheless, the shape of the curve remains essentially the same.

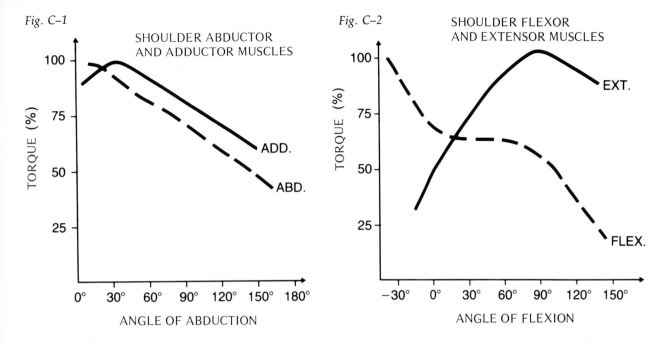

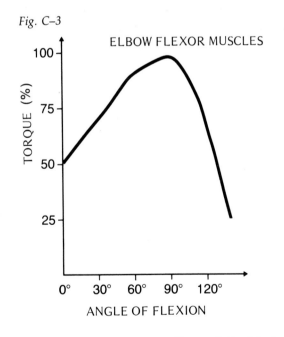

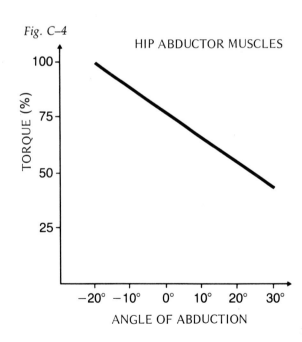

Figure C–1.   Maximum isometric torque at various joint angles for the shoulder abductor and adductor muscles. (From Bethe and Franke, 1919.)

Figure C–2.   Maximum isometric torque at various joint angles for the shoulder flexor and extensor muscles. (From Williams and Stutzman, 1956.)

Figure C–3.   Maximum isometric torque at various joint angles for the elbow flexor muscles. (From Williams and Stutzman, 1956.)

Figure C–4.   Maximum isometric torque at various joint angles for the hip abductor muscles. (From Olson et al., 1972.)

175

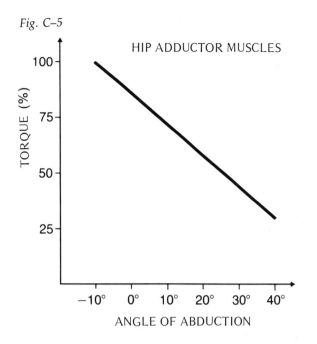

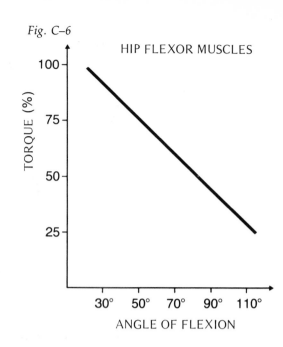

Figure C–5.   Maximum isometric torque at various joint angles for the hip adductor muscles. (From Williams and Stutzman, 1956.)

Figure C–6.   Maximum isometric torque at various joint angles for the hip flexor muscles. (From Williams and Stutzman, 1956.)

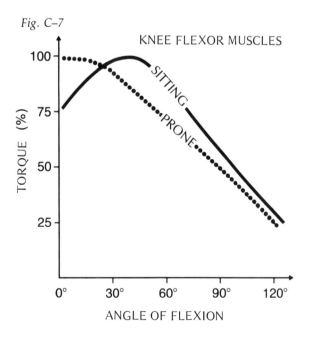

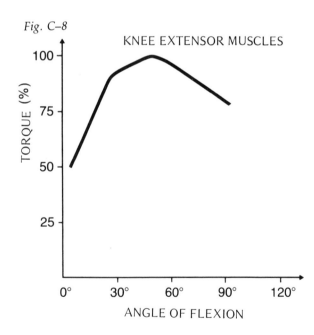

Figure C–7.   Maximum isometric torque at various joint angles for the knee flexor muscles with the subject in the sitting and prone positions. (From Williams and Stutzman, 1956.)

Figure C–8.   Maximum isometric torque at various joint angles for the knee extensor muscles. (From Haffajee et al., 1972.)

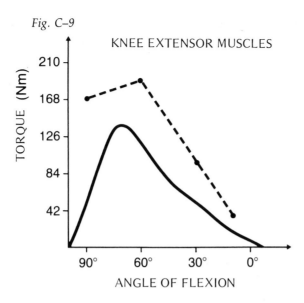

Fig. C–9

KNEE EXTENSOR MUSCLES

Figure C–9.   Maximum torque (Nm) produced by the knee extensor muscles at various angles of knee flexion for one person. The maximum isometric torque for 90°, 60°, and 10° of knee flexion is represented by the broken line. The isokinetic torque measured at an angular velocity of 90°/sec is represented by the solid line.

References

Aldman B.: Personal communication, Traffic Safety Institute, Chalmers Technical College, Gothenburg, Sweden, 1981.

Andersson G. B., Schultz A. B.: Transmission of moments across the elbow joint and the lumbar spine. J. Biomech. *12*:747–755, 1979.

Andersson G. B., Ortengren R., Herberts P.: Quantitative electromyographic studies of back muscle activity related to posture and loading. Orthop. Clin. North Am. *8*:85–96, 1977.

Andriacchi T. P.: Personal communication, 1981.

Andriacchi T. P., Goldflies J. O., Galante J. O.: Unpublished data, 1981.

Andriacchi T. P., Galante J. O., Belytschko T. B., Hampton S.: A stress analysis of the femoral stem in total hip prostheses. J. Bone Joint Surg. *58A*:618–624, 1976.

Andriacchi T. P., Andersson G. B., Fermier R. W., Stern D., Galante J. O.: A study of lower-limb mechanics during stair-climbing. J. Bone Joint Surg. *62A*:749–757, 1980.

Arborelius U. P., Ekholm J.: Mechanics of shoulder locomotor systems during exercises resisted by weight-and-pulley-circuit. Scand. J. Rehabil. Med. *10*:171–177, 1978.

Bandi, W.: Die retropatellaren Kniegelenkschäden. Pathomechanik und pathologische Anatomie, Klinik und Therapie. Aktuelle Probleme in Chirurgie und Orthopädie. Vol. 4. Bern, Hans Huber, 1977.

Bankov S., Jorgensen K.: Maximum strength of elbow flexors with pronated and supinated forearm. Communications from the Danish National Association for Infantile Paralysis, No. 29, 1969.

Bard G. B., Jones M. D.: Cineradiographic recording of traction of the cervical spine. Arch. Phys. Med. Rehabil. *45*:403–406, 1964.

Basmajian J. V.: *Muscles Alive. Their Functions Revealed by Electromyography.* 3rd ed. Baltimore, Williams & Wilkins, 1974.

Bethe A., Franke F.: Beiträge zum Problem der Willkürlich beweglichen Armprothesen. IV. Die Kraftkurven der indirekten natürlichen Energiequellen. München Med. Wschr. *66*:201–205, 1919. Cited in Brunnström S.: *Clinical Kinesiology.* 3rd ed. Philadelphia, F. A. Davis, 1972, p. 50.

Blount W. P.: Don't throw away the cane. J. Bone Joint Surg. *38A*:695–708, 1956.

British Orthopaedic Association: *Joint Motion. Method of Measuring and Recording.* Edinburgh, 1965.

Brodin H., Moritz U.: *Fysioterapi.* Lund, Sweden, Studentlitteratur, 1971.

Butler D. L., Noyes F. R., Grood E. S.: Ligamentous restraints to anterior-posterior drawer in the human knee. J. Bone Joint Surg. *62A*:259–270, 1980.

Cailliet R.: *Neck and Arm Pain.* Philadelphia, F. A. Davis, 1964.

Carlsöö S.: *How Man Moves. Kinesiological Studies and Methods.* London, William Heinemann, 1972.

Chao E. Y., Morrey B. F.: Three-dimensional rotation of the elbow. J. Biomech. *11*:57–73, 1978.

Colachis S. C., Strohm B. R.: A study of tractive forces and angle of pull on vertebral interspaces in the cervical spine. Arch. Phys. Med. *46*:820–830, 1965.

Colachis S. C., Strohm B. R.: Effect of duration of intermittent cervical traction on vertebral separation. Arch. Phys. Med. *47*:353–359, 1966.

Crowninshield R. D., Johnston R. C., Andrews J. G., Brand R. A.: The effects of walking velocity and age on hip kinematics and kinetics. Clin. Orthop. *132*:140–144, 1978.

Crue B. L., Todd E. M.: The importance of flexion in cervical halter traction. Bull. Los Angeles Neurol. Soc. *30*:95–98, 1965.

Dempster W. T.: Space requirements of the seated operator. Wright-Patterson Air Force Base, Ohio, WADCTR 55-159, 1955.

Denham R. A.: Hip mechanics. J. Bone Joint Surg. *41B*:550–557, 1959.

Denham R. A., Bishop R. E.: Mechanics of knee and problems in reconstructive surgery. J. Bone Joint Surg. *60B*:345–352, 1978.

Detenbeck L. C.: Function of the cruciate ligaments in knee stability. J. Sports Med. *2*:217–221, 1974.

Drillis R., Contini R., Bluestein M.: Body segment parameters. A survey of measurement techniques. Artif. Limbs *8*:44–66,1964.

Ekholm J., Arborelius U. P., Hillered L., Ortgvist A.: Shoulder muscle EMG and resisting moment during diagonal exercise movements resisted by weight-and-pulley-circuit. Scand. J. Rehabil. Med. *10*:179–185, 1978.

Farfan H. F.: *Mechanical Disorders of the Low Back.* Philadelphia, Lea & Febiger, 1973.

Floyd W. F., Silver P. H.: The function of the erector spinae muscles in certain movements and postures in man. J. Physiol. *129*:184–203, 1955.

Frankel V. H.: *The Femoral Neck: Function, Fracture Mechanisms, Internal Fixation.* Springfield, Charles C Thomas, 1960.

Frankel V. H.: Biomechanics of the hip. In *Surgery of the Hip Joint.* Edited by R. G. Tronzo. Philadephia, Lea & Febiger, 1973, pp. 105–125.

Frankel V. H.: Biomechanics of the hip. Lecture. Course in biomechanics. Swedish National Society of Certified Physiotherapists, Storlien, Sweden, 1979.

Frankel V. H., Burstein A.: *Orthopaedic Biomechanics*. Philadelphia, Lea & Febiger, 1970.

Frankel V. H., Nordin M.: *Basic Biomechanics of the Skeletal System*. Philadelphia, Lea & Febiger, 1980.

Friedenberg Z. B., Miller W. T.: Degenerative disc disease of the cervical spine. J. Bone Joint Surg. *45A*:1171–1178, 1963.

Goldie I. F., Reichmann S.: The biomechanical influence of traction on the cervical spine. Scand. J. Rehabil. Med. *9*:31–34, 1977.

Goodfellow J., Hungerford D. S., Zindel M.: Patello-femoral joint mechanics and pathology. I. Functional anatomy of the patello-femoral joint. J. Bone Joint Surg. *58B*:287–290, 1976.

Haffajee D., Moritz U., Svantesson G.: Isometric knee extension strength as a function of joint angle, muscle length and motor unit activity. Acta Orthop. Scand. *43*:138–147, 1972.

Harris P. R.: Cervical traction. Phys. Ther. *57*:910–914, 1977.

Inman V. T.: Functional aspects of the abductor muscles of the hip. J. Bone Joint Surg. *29*:607–619, 1947.

Johnston R. C., Larsson C. B.: Biomechanics of hip arthroplasty. Clin. Orthop. *66*:56–69, 1969.

Johnston R. C., Brand R. A., Crowninshield R. D.: Reconstruction of the hip. J. Bone Joint Surg. *61A*:639–652, 1979.

Keyserling W. M., Herrin G. D., Chaffin D. B.: Isometric strength testing as a means of controlling medical incidents on strenuous jobs. J. Occup. Med. *22*:332–336, 1980.

King A. I., Prasad P., Ewing C. L.: Mechanism of spinal injury due to caudocephalad acceleration. Orthop. Clin. North Am. *6*:19–31, 1975.

Lieb F., Perry J.: Quadriceps function. J. Bone Joint Surg. *50A*:1535–1548, 1968.

Lindahl O., Movin A.: Active-extension capacity of the knee-joint in the healthy subject. Acta Orthop. Scand. *39*:203–208, 1968.

Lindahl O., Movin A., Ringqvist I.: Knee extension. Acta Orthop. Scand. *40*:79–85, 1969.

Lindh M.: Increase of muscle strength from isometric quadriceps exercises at different knee angles. Scand. J. Rehabil. Med. *11*:33–36, 1979.

Lysell E.: Motion in the cervical spine. Acta Orthop. Scand., Suppl. *123*, 1969.

Markhede G.: Function-preserving surgery for extirpation of malignant soft-tissue tumors. A study of prognosis and function. Thesis. Departments of Orthopaedic Surgery, Pathology, and Rehabilitation Medicine, University of Gothenburg, Gothenburg, Sweden, 1980.

Markolf K. L., Graff-Radford A., Amstutz H. C.: In vivo knee stability. J. Bone Joint Surg. *60A*:664–674, 1978.

Markolf K. L., Mensch J. S., Amstutz H. C.: Stiffness and laxity of the knee—the contributions of the supporting structures. A quantitative in vitro study. J. Bone Joint Surg. *58A*:583–593, 1976.

Matthews L. S., Sonstegard D. A., Henke J. A.: Load bearing characteristics of the patello-femoral joint. Acta Orthop. Scand. *48*:511–516, 1977.

McLeish R. D., Charnley J.: Abduction forces in the one-legged stance. J. Biomech. *3*:191–209, 1970.

Merchant A. C.: Hip abductor muscle force. An experimental study of the influence of hip position with particular reference to rotation. J. Bone Joint Surg. *47A*:462–476, 1965.

Morris J. M.: Biomechanical aspects of the hip joint. Orthop. Clin. North Am. *2*:33–54, 1971.

Morrison J. B.: Bioengineering analysis of force actions transmitted by the knee joint. Bio-medical Engng. *3*:164–171, 1968.

Morrison J. B.: Function of the knee joint in various activities. Bio-medical Engng. *4*:573–580, 1969.

Morrison J. B.: The mechanics of the knee joint in relation to normal walking. J. Biomech. *3*:51–61, 1970.

Moritz U.: Traktionsbehandlingens mekaniska effekt på ryggraden. Sjukgymnasten *12*:19–22, 1975.

Murray M. P., Sepic S. B.: Maximum isometric torque of hip abductor and adductor muscles. Phys. Ther. *48*:1327–1335, 1968.

Murray M. P., Seireg A. A., Scholz R. C.: A survey of the time, magnitude and orientation of forces applied to walking sticks by disabled men. Am. J. Phys. Med. *48*:1–13, 1969.

Murray M. P., Guten G. N., Baldwin J. M., Gardner G. M.: A comparison of plantar flexion torque with and without the triceps surae. Acta Orthop. Scand. *47*:122–124, 1976.

Murray M. P., Baldwin J. M., Gardner G. M., Sepic S. B., Downs W. J.: Maximum isometric knee flexor and extensor muscle contraction. Phys. Ther. *57*:637–643, 1977.

Murray M. P., Guten G. N., Sepic S. B., Gardner G. M., Baldwin J. M.: Function of the triceps surae during gait. J. Bone Joint Surg. *60A*:473–476, 1978.

Nachemson A. L.: Lumbar interdiscal pressure. Acta Orthop. Scand., Suppl. *43*, 1960.

Nachemson A. L.: The influence of spinal movement on the lumbar intradiscal pressure and on the tensile stresses in the annulus fibrosus. Acta Orthop. Scand. *33*:183–195, 1963.

Nachemson A., Elfström G.: *Intravital Dynamic Pressure Measurements in Lumbar Discs: A Study of Common Movements, Maneuvers and Exercises*. Scand. J. Rehabil. Med., Suppl. 1. Stockholm, Almqvist & Wiksell, 1970.

Nicol A. C., Berme N., Paul J. P.: A biomechanical analysis of elbow joint function. In *Joint Replacement in the Upper Limb*. Conference sponsored by the Medical Engineering Section of the Institution of Mechanical Engineers and the British Orthopaedic Association, London, England, April 18–20, 1977. Mechanical Engineering Publications Ltd. for the Institution of Mechanical Engineers, London and New York, 1977, pp. 45–52.

Nissan M.: Review of some basic assumptions in knee biomechanics. J. Biomech. *13*:375–381, 1980.

Noyes F. R.: Functional properties of knee ligaments and alterations induced by immobilization. Clin. Orthop. *123*:210–242, 1977.

Olson V. L., Smidt G. L., Johnston R. C.: The maximum torque generated by the eccentric, isometric and concentric contractions of the hip abductor muscles. Phys. Ther. *52*:149–158, 1972.

Paul J. P., McGrouther D. A.: Forces transmitted at the hip and knee joint of normal and disabled persons during a range of activities. Acta Orthop. Belg. (Suppl.) *41*:78–88, 1975.

Perry J.: Normal upper extremity kinesiology. Phys. Ther. *58*:265–278, 1978.

Poppen N. K., Walker P. S.: Normal and abnormal motion of the shoulder. J. Bone Joint Surg. *58A*:195–201, 1976.

Poppen N. K., Walker P. S.: Forces at the gleno-humeral joint in shoulder abduction in the scapular plane. Clin. Orthop. *135*:165–170, 1978.

Ramsey P. L., Hamilton W.: Changes in tibiotalar area of contact caused by lateral talar shift. J. Bone Joint Surg. *58A*:356, 1976.

Reilly D., Martens M.: Experimental analysis of the quadriceps muscle force and patello-femoral joint reaction force for various activities. Acta Orthop. Scand. *43*:126–137, 1972.

Rolander S. D.: Motion of the lumbar spine with special reference to the stabilizing effect of posterior fusion. An experimental study on autopsy specimens. Acta Orthop. Scand., Suppl. *90*, 1966.

Rydell N. W.: Forces acting on the femoral head-prosthesis: A study on strain gauge supplied prostheses in living persons. Thesis. Department of Orthopaedic Surgery, University of Gothenburg, Gothenburg, Sweden, 1966.

Sammarco G. J., Burstein A. H., Frankel V. H.: Biomechanics of the ankle: A kinematic study. Orthop. Clin. North Am. *4*:75–96, 1973.

Schlicke L. H., White A. A., Panjabi M. H., Pratt A., Kier L.: A quantitative study of vertebral displacement and angulation in the normal cervical spine under axial load. Clin. Orthop. *140*:47–49, 1979.

Schultz A. B., Andersson G. B.: Analysis of loads on the lumbar spine. Spine *6*:76–82, 1981.

Schultz A. B., Andersson G. B., Ortengren R., Björk R., Nordin M.: Analysis and quantitative myoelectric measurements of loads on the lumbar spine when holding weights in standing postures. Spine *7*:390–397, 1982a.

Schultz A. B., Andersson G. B., Ortengren R., Haderspeck K., Nachemson A. L.: Loads on the lumbar spine. Validation of a biomechanical analysis by measurements of intradiscal pressures and myoelectric signals. J. Bone Joint Surg. *64A*:713–720, 1982b.

Schultz A. B., Andersson G. B., Haderspeck K., Ortengren R., Nordin M., Björk R.: Analysis and measurement of lumbar trunk loads in tasks involving bends and twists. J. Biomech. *15*:669–675, 1982c.

Scudder G. N.: Torque curves produced at the knee during isometric and isokinetic exercise. Arch. Phys. Med. Rehabil. *61*:68–73, 1980.

Seireg A., Arvikar R. J.: The prediction of muscular load sharing and joint forces in the lower extremities during walking. J. Biomech. *8*:89–102, 1975.

Shenkin H. A.: Motorized intermittent traction for treatment of herniated cervical disk. J.A.M.A. *156*:1067–1070, 1954.

Smidt G. L.: Biomechanical analysis of knee flexion and extension. J. Biomech. *6*:79–92, 1973.

Smith J. W.: The forces operating at the human ankle joint during standing. J. Anat. *91*:545–564, 1957.

Stauffer R. N., Chao E. Y., Brewster R. C.: Force and motion analysis of the normal diseased and prosthetic ankle joint. Clin. Orthop. *127*:189–196, 1977.

Stoddard A.: Traction for cervical nerve root irritation. Physiotherapy *40*:48–49, 1954.

Sutherland D. H.: An electromyographic study of the plantar flexors of the ankle in normal walking on the level. J. Bone Joint Surg. *48A*:66–71, 1966.

Svedenkrans M.: *Arbete med och utan underarmsstöd. En studie av belastning på nacke och axlar.* Järfälla, Sweden, IBM Svenska AB, 1980.

Swedish National Board of Safety and Health: *Lyft och bärarbete. Fysiologiska och medicinska effekter på människan.* Discussion by B. Nordgren. Stockholm, 1980, p. 269.

Walker L. B., Harris E. E., Pontius U. R.: Mass, volume, center of mass, and mass moment of inertia of head and head and neck of human body. Proceedings of the Seventeenth Stapp Car Crash Conference, Society of Automotive Engineering, Inc. Two Pennsylvania Plaza, New York, 1973.

Wahrenberg H., Lindbeck L., Ekholm J.: Knee muscular moment, tendon tension force and EMG during a vigorous movement in man. Scand. J. Rehabil. Med. *10*:99–106, 1978.

White A. A., Panjabi M. M.: *Clinical Biomechanics of the Spine.* Philadelphia, J. B. Lippincott, 1978.

Williams M., Lissner H.: *Biomechanics of Human Motion.* Edited by B. LeVeau, 2nd ed. Philadelphia, Saunders, 1977.

Williams M., Stutzman L.: Strength variation through the range of joint motion. Phys. Ther. Rev. *39*:145–152, 1956.

Youm Y., Dryer R. F.: Biomechanical analysis of forearm pronation-supination and elbow flexion-extension. J. Biomech. *12*:245–255, 1979.

Zernicke R. F., Garhammar J., Jobe F. W.: Human patellar-tendon rupture. J. Bone Joint Surg. *59A*:179–183, 1977.